Collecting Military Headgear

COLLECTING MILITARY HEADGEAR

A GUIDE TO 5000 YEARS OF HELMET HISTORY

Robert Attard

Schiffer Military History
Atglen, PA

Book Design by Ian Robertson.

Library of Congress Control Number: 2004109259

Printed in China.
ISBN: 0-7643-2102-1

Published by Schiffer Publishing Ltd.
4880 Lower Valley Road
Atglen, PA 19310
Phone: (610) 593-1777
FAX: (610) 593-2002
E-mail: Info@schifferbooks.com.
Visit our web site at: www.schifferbooks.com
Please write for a free catalog.
This book may be purchased from the publisher.
Please include $3.95 postage.
Try your bookstore first.

In Europe, Schiffer books are distributed by:
Bushwood Books
6 Marksbury Avenue
Kew Gardens
Surrey TW9 4JF
England
Phone: 44 (0) 20 8392-8585
FAX: 44 (0) 20 8392-9876
E-mail: Info@bushwoodbooks.co.uk.
Free postage in the UK. Europe: air mail at cost.
Try your bookstore first.

CONTENTS

Acknowledgments

This book would not have been possible without the generous assistance of a number of friends all around the world.

I wish to thank Dr. Jean-Philippe Morange Ph.D, Multidisciplinary Analyst at the United Nations for providing me with photos of fine Napoleonic and Louis Phillipe period headgear from his impressive collection of French 19th military accoutrements. I also wish to thank Vincent of Gallery Samurai, Mukashiya K.K of Tokyo, Japan for providing me with valuable assistance regarding the Kabuto and providing me with pictures of valuable Japanese helmets. I am also grateful to Mr. Stefan Anbratt from Sweden for providing me with photos of an exceptional item from his advanced collection. A special word of thanks also goes to Mr. Johan Somers, a particularly knowledgeable and friendly Belgian collector from Wommelgem who exercised his best efforts to provide me with a considerable number of first hand photos of rare Prussian, Belgian and Bavarian headgear from the continent. A special word of thanks goes to Dr. Ernst Wagner of Hermann Historica of Munich who kindly provided me with photographs of some truly impressive classic Greek and roman helmets as well as some exceptionally rare 18th century headgear.

I am most grateful to Martin Bonnici, my promising, young Maltese photographer who diligently photographed the helmets from Malta and provided me with his patient advice regarding illustrations. I must also thank Romina Maria for typing the book and Mr. Ian Robertson, my editor at Schiffer books for his assistance throughout the whole publication process. A word of thanks also goes to Fondazzjoni Patrimonju Malti for allowing me to use photos of helmets from the collection at Palazzo Falson and Mr. Tom Berndt of the Militaria International Magazine for sending me photographs of fine helmets from the MAX Show.

I also wish to thank Dr. Edward De Bono, the world's leading authority in the direct teaching of thinking as a skill and the Edward Debono Foundation (Malta) for being the first publisher that entrusted me with my first publication (the Malta, A Collection of Tales & Narratives series). Lastly I wish to thank my family, especially my wife, for her constant support in my collecting interest.

"So spake glorious Hector, and stretched out his arm to his boy. But the child shrunk crying to the bosom of his fair girdled nurse, dismayed at his dear father's aspect, and in dread at the bronze and horse hair crest that he beheld nodding fiercely from the helmet's top. Then his dear father laughed aloudÖforthwith glorious Hector took the helmet from is head, and laid it, all gleaming, upon the earth, then kissed he his dear son and dandled him in his arms, and spake in prayer to Zeus and all the Gods, 'O Zeus and all ye gods, vouchsafe ye that this my son may likewise prove even as I, Pre-emminent amid the Trojans, and as valiant in might, and be a great king of Ilios."

From Homer's Iliad
Circa 8th century B.C.
(Wordsworth Classics)

An exceptional M 1852 Russian Officer's Lifeguard Horse Guard Regiment's helmet c. 1860. *(Courtesy of Hermann Historica)*

Side view of the M 1852 helmet. *(Courtesy of Hermann Historica)*

Close up view of the M 1852 helmet's finial. *(Courtesy of Hermann Historica)*

A truly extraordinary Greek suit of armour from the 5th century BC. *(Courtesy of Hermann Historica)*

A Somalian tooled leather shield c. 1890. The construction techniques used in the production of this heavy shield reflect an ancient craftsmanship that could have been used in the production of helmets. *(Private Collection - Malta)* Photo by M Bonnici

In the 1960s Professor Andrew Snodgrass[7] identified the following types of Greek helmets:

The Pilos Helmet;
The Boeotian helmet;
The Corinthian Helmet
The Chalcidian Helmet;
The Illyrian Helmet.

The Pilos helmet was a run of the mill helmet of very simple construction and form. It consisted of an elongated one-piece cone. It was used by infantrymen, and is probably the oldest form of headgear ever to appear on the military scene. It is the most common and cheapest form of ancient helmet. Various forms of this early headgear survive in private collections and museums. A particularly interesting pilos helmet of a quasi-identical form to that of a 16th century Spanish Morion is kept at the *Musee' de l'Armee* in Paris. A leading German auctioneer[8] auctioned some fine pilos helmets in May 2003. All these helmets came from the prestigious Axel Guttmann[9] Collection and were valued in the region of USD 7,000.

The Boeotian helmet was a cavalry helmet that consisted of a conical skull and wide brim projecting over the eyes, ears, and nape.[10] It is one of the rarest of all ancient helmets, and the author never had the opportunity to notice a Boetian helmet being offered for sale. The Corinthian helmet is the Greek helmet that is in the highest demand. It is the Greek helmet that is generally associated with Alexander the Great and the Hellenistic age. The Corinthian helmet was one of the first closed helmets of history. It consisted of a bronze dome that extended itself over the nose and cheeks, enveloping the wearer's face almost completely.[11] Plutarch was probably describing a Corinthian helmet when he wrote:

"*But after Masistius' horse received a wound, and flung him, and he falling could hardly raise himself through the weight of his armor, the Athenians, pressing upon him with blows, could not easily get at his person, armed as he was, his breast, his*

A Macedonian Pilos helmet, 4th century BC. *(Courtesy of Hermann Historica)*

A Roman iron mask helmet of Silistra type (1st to mid 2nd century AD). *(Courtesy of Hermann Historica)*

A Roman iron mask helmet of Silistra type (1st to mid 2nd century AD). *(Courtesy of Hermann Historica)*

A front view of the Silistra helmet. *(Courtesy of Hermann Historica)*

A Roman iron mask helmet of Resca type c. 2nd century AD. *(Courtesy of Hermann Historica)*

Side view of the Resca helmet. *(Courtesy of Hermann Historica)*

De Cosson's *Catalogue of the Exhibition of Ancient Helmets and Examples of Mail*. In the nineteenth century Italian craftsmen produced replicas of roman helmets and late renaissance helmets that may easily confuse inexperienced collectors. These early replicas are collectibles in their own right and sell for considerable prices.[23]

Recent blockbusters, such as *Gladiator*, led modern day armourers to produce replica *Silistra* and *Mormilliones* helmets; these replicas would fool few collectors, however, because the standard of workmanship is not comparable to that of the armourers of ancient Rome.

Medieval Helmets

1.3.1 The *Spangenhelm*

"Tristram had not ridden far when he met a damsel, who cried out, 'Ah, my lord! hasten forward, and prevent a horrid treason!' Tristram flew to her assistance, and soon reached a spot where he beheld a knight, whom three others had borne to the ground, and were unlacing his helmet in order to cut off his head."

Thomas Bulfinch: '*King Arthur and his Knights Tristan and Isolde*

The helmets worn in the early middle ages were very similar to those worn in the final days of the Roman Empire. At times it is difficult to distinguish these helmets from late Roman helmets.[24] One of the helmets characteristic of the early middle ages was the *Spangenhelm*, a hemispherical helmet constructed from overlapping iron plates. *Spangenhelms* were worn by Carolingian troops, Ostrogothic troops, the Normans, the Franks, and the Vikings. *Spangenhelm* fell into disuse in Western Europe only in the 14th century.[25] *Spangenhelms* may be observed in various museums, and appear on an almost regular basis in the advanced auction houses.[26] An interesting collection of early medieval Viking helmets is kept at the archeological museum in Dublin. Early medieval helmets are very rare, and even heavily rusted, dredged, or excavated examples tend to sell for over USD 1,000. The author recently noticed a particularly curious *Spangenhelm* being offered for sale. It was heavily corroded, and less than 50% of the original helmet remained. However, this helmet was 'complete' with an odd accessory—the deformed skull of its wearer!

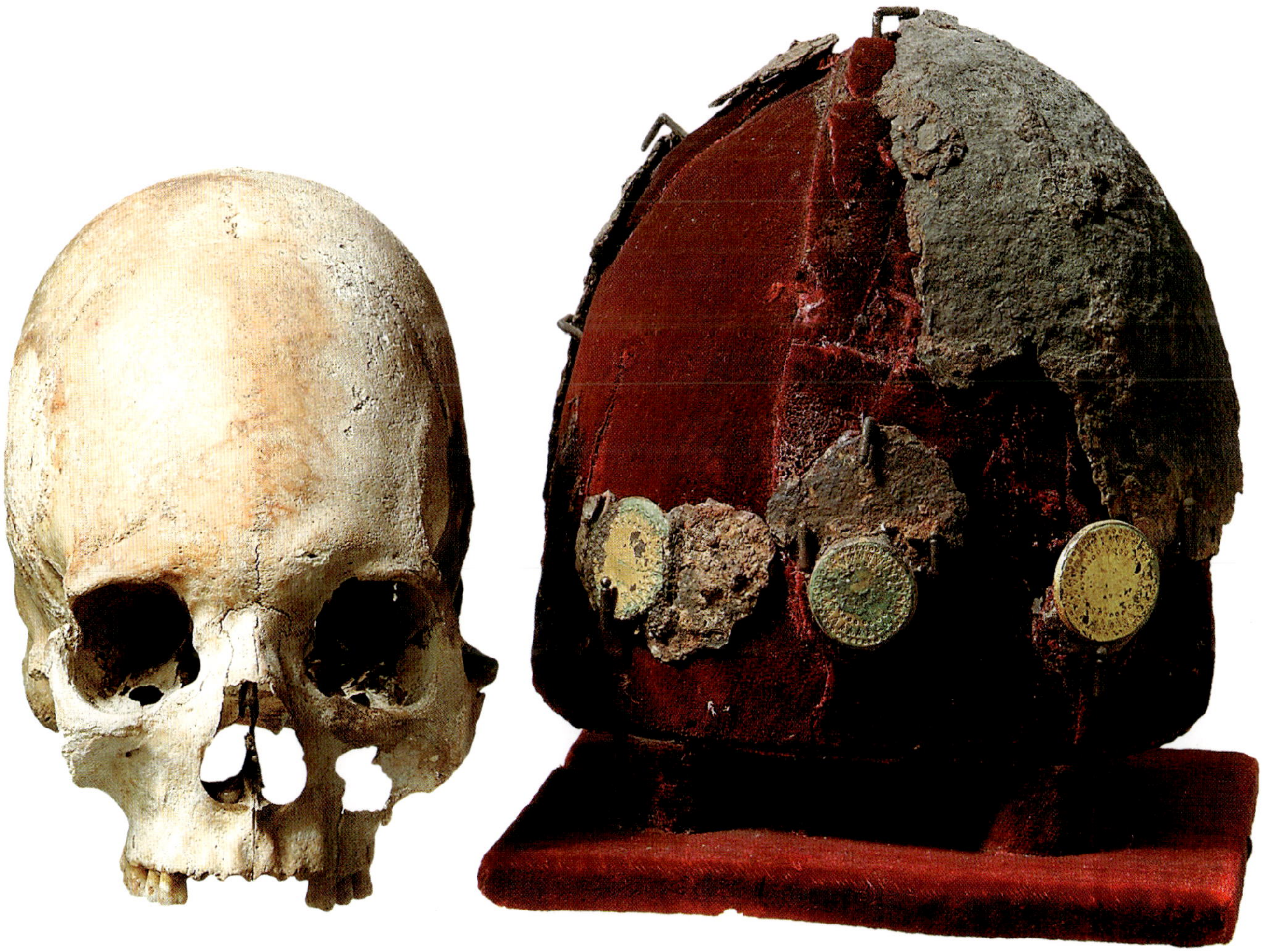

The most odd helmet noticed by the author, an excavated spangenhelm from the dark ages with its wearer's skull en suite. *(Courtesy of Hermann Historica)*

Side view of the excavated spangenhelm. *(Courtesy of Hermann Historica)*

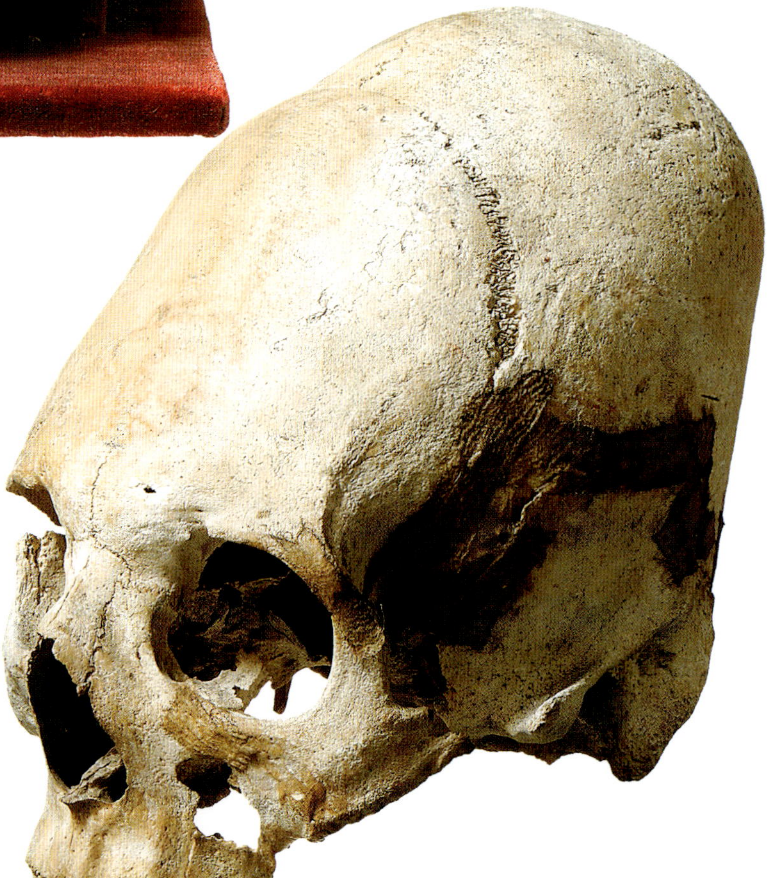

This most unusually shaped human head was artificially deformed during its owner's lifetime. *(Courtesy of Hermann Historica)*

1.3.2 Conical Helmets and Great Helms

"I saw him down; thrice up again, and fighting;
From helmet to the spur all blood he was."

From Shakespeare's Henry the Fifth
Act IV Scene VI

The conical helmet with nasal bar, characteristic of the Normans worn in the 11th, 12th, and 13th centuries,[27] portrayed in the famous *Bayeux* tapestry is another interesting helmet from this period, but the author has never seen a conical helmet with nasal bar being offered for sale. Examples of this rare helmet survive in some continental museums. One of the most celebrated examples is probably the gashed helmet of Saint Wenceslaus in the Prague Cathedral dating back to the tenth century.

The late 12th century saw the appearance of a helmet that became one of the most fitting symbols of the age of chivalry, the Great Helm.[28] In the first quarter of the 13th century this helmet developed in the large barrel shaped helmet characteristic of the middle ages. These impressive helmets were often embellished with colourful and grotesque[29] crests made of *gesso* or boiled leather.[30] The multi-plate Great Helmet was worn in battle up to the fourteenth century, but remained in use as a jousting helmet and in funerary effigies up to the sixteenth century.[31]

Great Helms were often used to adorn the tombs of Knights and stood the test of time. The Great Helm of Sir Richard Pembridge and the Helm of the Great Helm of the Black Prince at Canterbury Cathedral are two of the most frequently photographed Great Helms. A handful of advanced private collectors own Great helms in their collections. An advanced dealer was offering a quasi-identical Great Helm to the Great Helm represented in a sculpture in Wells Cathedral for sale around five years ago.[32] It weighed eight Libras and was just over eleven inches tall. It was in excellent condition for the age.[33] The price was not disclosed in the catalogue. Fake Great Helms were being produced in the mid-nineteenth century. The British Archeological Association exhibited some of these fakes in 1852.[34]

Sallets and Bascinets

"She saw the helmet and the plume,
She looked down to Camelot.
Out flew the web and floated wide;
The mirror cracked from side to side;
'The curse is come upon me,' cried
The Lady of Shalott"

Alfred Tennyson (1809-1892)
The Lady of Shalott

In the first years of the 14th century the conical helmet developed into a helmet known as "the bascinet." The bascinet was a conical helmet that extended itself to cover the nape of the neck and ears of the wearer.[35] The parts of the head and neck of the wearer that were left exposed by the plate of the bascinet were defended by links of riveted mail. It appears that towards the late 14th century the bascinet started to be fitted with face guards. In the earlier versions the face guard seems to have consisted of a simple, globosely shaped mask with eye slots, while after the mid-14th century the face guard took the shape of a pointed snout.[36] Bascinets with the pointed snout were known as "hounskulls," and they represent one of the most historically important forms of medieval headgear. One of the finest hounskulls produced in the fourteenth century is an example held in the"Musee' de Cluny.[37] In the first quarter of the 15th century the closed bascinet developed into†"the Great Bascinet," a heavier bascinet with a rounded visor.

At this time another helmet, the armet, developed from the bascinet. It was another multi-plate helmet that consisted of a skull; a pair of hinged face pieces fastened to the front, a brow plate, and a visor. In the sixteenth century the armet developed into the closed helmet—a full-face helmet of a more simple construction.

This helmet remained in use up to the early seventeenth century when it evolved into odd variants, such as the Savoyard and the curassier's helmet with caged visor.

The term '*sallet*' is often used to denote early bascinets, as well as two 'families of helmets' that developed from this early

Spangenhelm
6th Century A.D

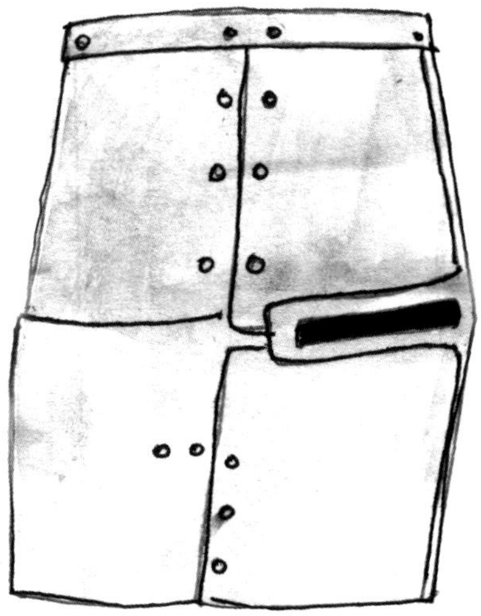

Great Helm
13th century A.D

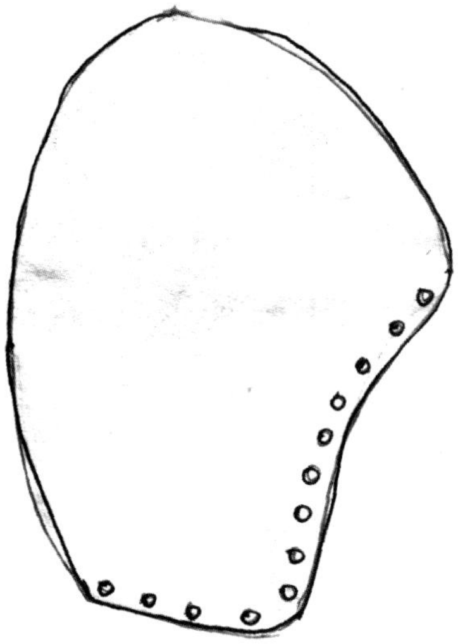

Sallet
c.1370 A.D

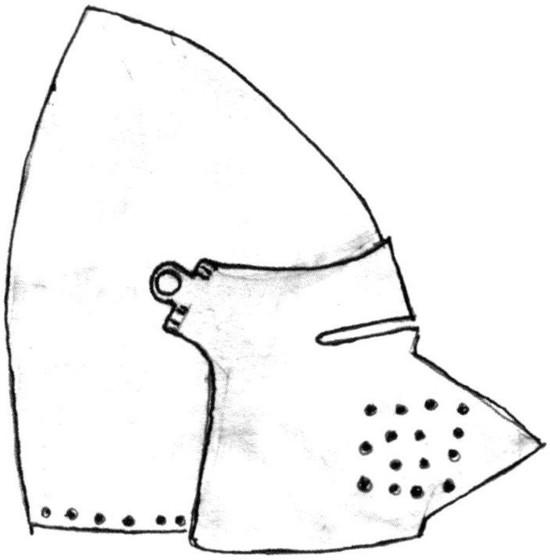

Hounskull
c. 1390 A.D

Kettle Hat
Late 15th century A.D

Artist's impression of some medieval helmets.

A late 16th century close helmet. *(Courtesy of Fondazzjoni Patrimonju Malti)*

helmet of simple form. Italy and Germany devised various forms of sallet. One of the Italian variants consisted of a one-piece helmet of simple construction that revived the shape of the ancient Corinthian helmet and was known as the barbuta.

The Italians developed sallets of a more complex construction. The latter consisted of a skull piece of bascinet form, a visor, and brow plate. One of the finest helmets of this variety is the 'Malta Sallet' at the Palace Amouries Museum in Malta. It is a Venetian sallet from the renowned Missaglia workshop. It dates circa 1490-1520 and was identified by one of the greatest connoisseurs of all time, the late Sir Francis Laking.[38]

Various sallet patterns were created in Germany. One of these helmets consisted of a one-piece helmet that completely occulted the face. Vision was provided by a narrow slit for the eyes. The Germans in the First World War wore quasi-identical helmets to these sallets. Another version of the sallet used in Germany in the 15th century was fitted with a lobster-tail—another important feature in the evolution of military headgear[39].

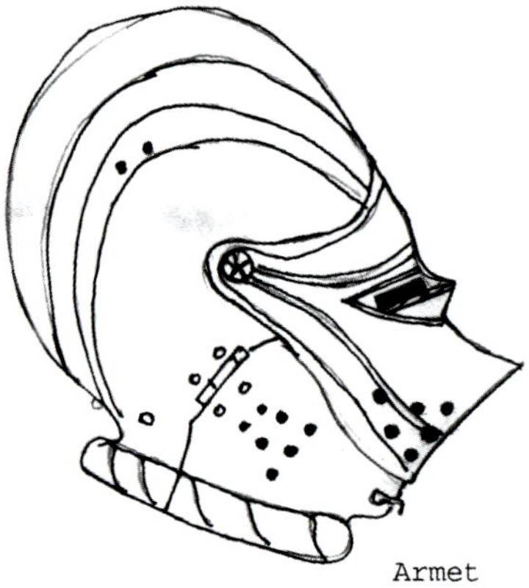

Armet
c. 1520-1530 A.D

Close Helmet
c.1539 A.D

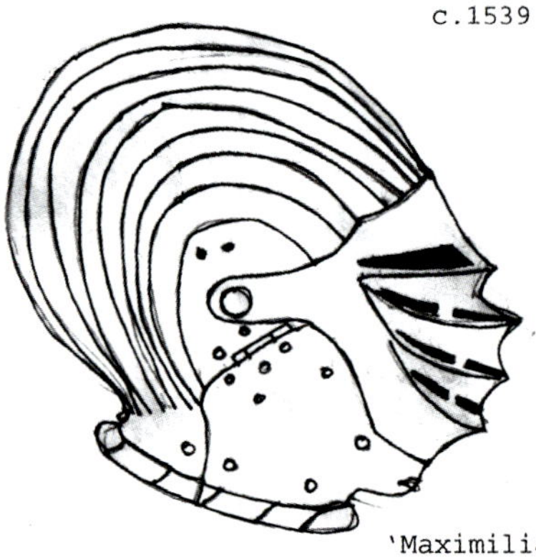

'Maximilian' Armet
c. 1520 A.D

Artist's impression of armets and close helmets.

A French Close helmet c 1590. *(Courtesy of Hermann Historica)*

Side view of the close helmet. *(Courtesy of Hermann Historica)*

An Italian sallet of Barbuta form from the 15th century. *(Courtesy of Hermann Historica)*

An Italian sallet of Barbuta form from the 15th century. *(Courtesy of Hermann Historica)*

A considerable number of bascinets appeared on the market in the first half of 2003. A relic example of simple construction was valued at USD 1,000 by Mohawk Arms, while two other one piece relic 14th century bascinets were sold on ebay. One of these bascinets was offered with buy it now facility at USD 6,240,[40] while a second example was sold at USD 795.[41]

Close helmets and the earlier armets tend to sell within the USD 3,000 to the USD 10,000 price range. Needless to say, sallets and close helmets have been extensively faked. In 1922 Laking[42] described a replica bascinet that was applied with '*composition resembling rust*.' Some of these fakes were purchased by museum curators and displayed as genuine pieces by renowned museums, such as the Museé d'Artillerie and the Tower of London.[43] Some of the Victorian replicas tend to be exceptionally deceptive on account of their age, and collectors should be very careful.

Notes:

[1] Becker Edward. B: Helmets Developments & Standards;
[2] These include coins, bronzes and friezes;
[3] The early authors that refer to helmets include Herodotus, Homer and Virgil.
[4] The resilience of these materials cannot be overlooked. The author had the opportunity to inspect African shields made from dried ox hides, cowhides and other animal skins and was impressed by the resilience of these materials;
[5] Gurdijan E.S: Head Injury from Antiquity to the Present with Special Reference to Penetrating Head Wounds;
[6] Homer ibid Book VI;
[7] Snodgrass A.M: Arms and Armour of the Greeks;
[8] Hermann Historica May 2003 sale;
[9] The late Mr. Guttman was a leading Jewish Banker who fell victim to the Nazi Holocaust;
[10] Bull Stephen: An Historical Guide to Arms and Armour;
[11] A quasi-identical helmet to the Corinthian helmet, the Barbuta was worn in Continental Europe in the 15th century vide infra;
[12] Plutarch: Plutarch's Lives (Aristides);
[13] It was almost impeccable in terms of construction and ageing. Only a close look at the helmet with a magnifying glass revealed some flaws minor flaws in construction (casting). It was being offered for the price of USD 1,000;
[14] Bull ibid;
[15] Planche': Cyclopedia of Costume;
[16] Infantymen
[17] Bull ibid pp.31;
[18] Tacitus: Germany;
[19] Vide Cap 3 Infra;
[20] Crests;
[21] This helmet was excavated in the 18th century. Vide Bull op. cit pp. 34;
[22] ibid Hermann Historica;
[23] Frequently prices for these early helmets exceed USD 1,000;
[24] *ibid* Hermann Historica Lots 16, 17 and 108;
[25] Blaire Claude: European Armour pg. 25
[26] Including the Italian Auctioneer Casa D'Aste Czerny;
[27] Blair ibid pp. 25-26;
[28] Blair ibid pp. 30 et seq;
[29] Illuminated Manuscript at the Biblioteque Nationale Paris published in Ullman Walter (Editor): Medieval Europe pg. 37;
[30] Edge David & Miles Paddock: Arms and Armour of the Medeival Knight pg. 71;
[31] Attard Robert: The Great Helm (The Armourer Magazine Issue 20) pp. 15 and 16;
[32] Peter Finer Catalogue Lot 3;
[33] This helmet dated circa 1300;
[34] Laking Guy Francis: A record of European Armour Ö Vol V p. 115
[35] Blaire ibid pg.51;
[36] Blaire ibid figures 68-74;
[37] Illustrated in Ullmann ibid pg.75 Vol 1;
[38] Laking Guy Francis; A Catalogue of the Armour and Arms in the Armoury of the Knights of Saint John of Jerusalem;
[39] Blaire ibid figures 103-115;
[40] Item#2169617429;
[41] 21771110648;
[42] Laking op. cit. Vol V p 115;
[43] Laking op cit Vol V p 114;

2

Morions and Burgonets (1530-1620)

From Lepanto to the New World

"Good sirs,' quoth the barber, 'what do you think of that which is affirmed by these gentlemen, who yet contend that this is not a basin, but a helmet?" 'He that denies it,' quoth Don Quixote, 'I will make him know that he lies, if he be a knight; and if he be but a squire, that he lies and lies again a thousand times.' Our barber, who was also present, as one that knew Don Quixote's humour very well, would fortify his folly and make the jest pass yet a little further, to the end that they all might laugh; and therefore, speaking to the other barber, he said,' 'Sir barber, or what else you please, know that I am also of your occupation, and have had my writ of examination and approbation in that trade more than these thirty years, and am one that knows very well all the instruments of barbery whatsoever; and have been besides, in my youthful days, a soldier; and do therefore likewise know what is a helmet, and what a morion, and what a close castle....'"

Miguel de Cervantes Saavedra
Don Quixote
1605 AD

The helmets worn in the sixteenth century may be classified in the following categories:

1) Open faced Helmets (Morions and late Kettle-hats);
2) Semi-Open Faced Helmets (Burgonets);
3) Full Faced Helmets (Cuirassier helmets and Savoyards)

All three forms of headgear saw extensive use in 16th century, but full-faced helmets were very uncomfortable and had to give way to open-faced and semi-open faced helmets. In the 17th century only chosen men, such as siege engineers, cuirassiers, and commanders used full-faced helmets. The rest of the troops wore practical and lightweight open-faced and semi-open faced helmets.[1] By the end of the 17th century all forms of metal headgear almost fell completely into disuse.

2.1 Open Faced Helmets

The kettle-hat (an iron helmet consisting of a hemispherical bowl with a wide brim) was one of the most successful helmets in military history. It was in use in the ancient world, was extremely popular throughout the Middle Ages,[2] and after around two centuries of disuse returned on the battlefields of World War One[3] and World War Two.[4]

The 16th century provided an ideal setting for a rise in the popularity of the kettle hat. The 16th century was a period of transition in the history of Armour. In the 16th century the cavalry wore closed helmets reminiscent of the Middle Ages, while the infantry was equipped with the forerunners of 20th century ballistic helmets. Improvements in the production of firearms and their widespread use meant that the days of full armour and close helmets were numbered. Arquebusiers—infantrymen equipped with firearms—needed a helmet that afforded their wearers a wide angle of vision, and the kettle-hat was a helmet that satisfied this requirement.

An early morion from the 1530s. *(Courtesy of Hermann Historica)*

Side view of the early morion. *(Courtesy of Hermann Historica)*

In the 16th century the kettle-hat was no longer referred to as a 'kettle hat,' but was instead referred to as a 'Morion,'[5] a word that is believed to be derived from the Italian word '*Moro,*' meaning Moor.

Evidence of the effectiveness of the Morion in battle is contained in the diary of a 16th century Spanish mercenary named Balbi Da Correggio.[6] Da Correggio refers to the special ball-proof morions that were produced in his own days to counter-react to the threat posed by the firearm. He held these morions in very high repute:

"By mischance (Robles) was not wearing his bullet proof helmet. Never did he require it more, for, while looking downwards from the wall, a shot was aimed at him from the enemy trenches which hit him in the temple and he fell without uttering a word."[7]

Although ball proof morions may have been a formidable defence versus arquebuse fire:

"While preparing to go he was hit by a shot from an arquebuse which dented his helmet and he fell."[8]

unfortunately, even these helmets were of no help against artillery. The same source narrates:

"The ensign of Romegas was killed and Admiral Dural was wounded by a slug from a basilisk which penetrated his helmet."[9]

Da Correggio also refers to the custom of decorating helmets with feather plumes.[10] In fact, feather-hackles were often noticed on Morions:[11]

"6 August 1565 On the 6th during the dinner hour with his arquebuse on his shoulders, and wearing his helmet with many feather he went to the Post of Provence."

The swept peaks of the Peaked Morion made it suitable for use by soldiers who required a wide angle of vision. (Private Collection - Malta) Photo by M Bonnici

Morions were worn in conjunction with plate cheek pieces.[12] These leather straps, covered with 3 to 4 metal scales, supplied a limited degree of face protection.

A further accessory provided additional protection to the exposed face. This was the beavor, or buff,[13] a lamellar cup protecting the lower face. This accessory converted the morion into a closed helmet.

Morions appear to have been fitted with a canvas lining system[14] similar to the linings used on some early nineteenth century Shakos.[15] However, the author did not notice a morion complete with its lining being offered for sale. It appears that in some Comb Morions the lining used to consist of a quilted hat worn under the helmet,[16] similar to the suspension system adopted on the USM1 helmet from mid-World War II to the 1980s.

Various patterns of Morions were used in Europe, and 16th century Morions may be classified in three categories:

a) The Peaked Morion;[17]
b) The 'Spanish' Morion;[18]
c) The Comb Morion.[19]

2.1.1 The Peaked Morion

"His helmet now shall make a hive for bees...."

George Peel
A Farewell to Arms
C.1590 AD

The Peaked Morion is probably the rarest type of Morion. It was very popular with arquebusiers and archers on account of the fact that it had swept peaks that afforded its wearer a wide angle of vision. Its simple almond shaped skull, identical to the almond shaped skull of the 'Spanish' Morion, facilitated its mass production.

The earliest Peaked Morions with their characteristic pear stalk finial date back to circa 1530.[20] At this point in time the

Side view of an Italian Morion from the second half of the sixteenth century. This helmet was published in Antique Collecting Vol 36. No. 8. Photo by M Bonnici

The peaked Morion had a very similar shape to the 'Spanish Morion.' The swept up peaks were the only points of difference. Photo by M Bonnici

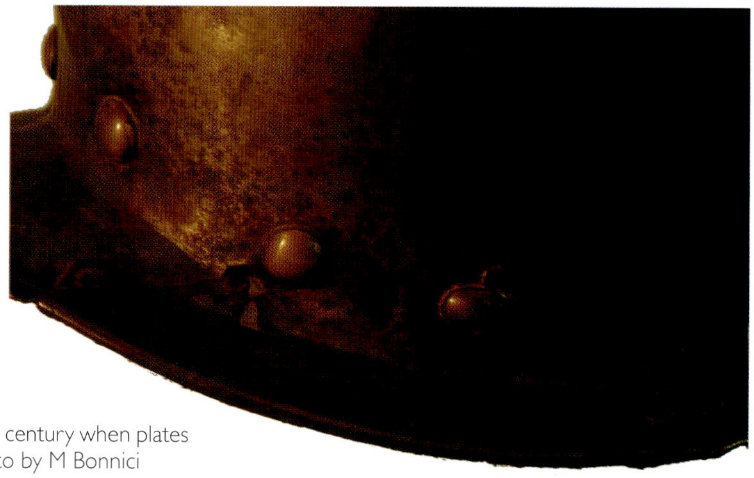

This peaked Morion appears to have been restored in the 19th century when plates were riveted onto the helmet to cover up missing parts. Photo by M Bonnici

peaked morion retained the wide brim characteristic of the kettle-hat. This brim shrunk as time went by, and by the 1570s it was only around two inches thick. On the other hand, the peaked morions produced in the late 16th century had taller and more slender skulls to those produced in the 1530s. Some peaked morions were embellished with fine gilding and delicate etching.[21] The finest examples were decorated with embossing. Most peak morions were forged in one piece, while some very rare examples were forged in two pieces. The heights of peaked Morions span from around eight inches in the earlier examples to around twelve inches in the later examples.

2.1.3 The 'Spanish' Morion

"He armed himself cap-a-pie, laced on his ill-contrived helmet, braced on his target, grasped his lance, mounted Rosinante...."

Miguel de Cervantes Saavedra
Don Quixote
1605 AD

The Morion known as the 'Spanish' Morion is probably the commonest form of Morion. The shape of the skull is the same almond shaped skull with pear stalk finial as the peaked morion. The shape of the brim is the only point of difference;

A plain 'Spanish' Morion dating c. 1600 of a very small size. (Private collection - Malta) Photo by M Bonnici

Inner view of the morion. Photo by M Bonnici

Side view of the morion. Photo by M Bonnici

in the 'Spanish' Morion the brim is horizontal. At times it is very difficult to fit certain helmets within the 'Peaked' or the 'Spanish' categories on account of the fact that in some helmets the brims are only slightly swept up.

The traditional 'Spanish' morion probably came into being roughly at the same time as the peaked morion. However, Blaire traces the origins of both types of helmets to the late 15th century.[22] It should be noted that the term 'Spanish' Morion is actually a misnomer. 'Spanish' Morions were not only produced in Spain, but also in other countries, such as Italy. The Spanish Morion remained in use up to the first quarter of the 17th century,[23] when it was substituted by a similar helmet.[24] Almost all the surviving 'Spanish' Morions were formed in one piece. The author noticed 'Spanish' morions as low as seven inches and as high as twelve inches. In this type of helmet height cannot be probably taken as a dating feature. As with the peaked morions, some 'Spanish' morions were embellished with gilding, embossing,[25] or etching. Examples made of tooled leather exist in museums,[26] however, the author never noticed leather examples being offered for sale.

It is interesting to note that 'Spanish' morions were not only extensively used all around Europe, but that some 'Spanish' Morions were also used and produced as far away as Japan. The Japanese used to import and make their own 'Spanish Morions,' which they used to call' *'Namban Hachi.'*[27] The Japanese Samurai kept wearing 'Spanish' Morions up to the 18th century, a time when the Spanish Morion fell in disuse in Europe.[28] Once the author noticed an illustration of a Japanese suit of Armour with a late 16th century Spanish Morion of Italian manufacture, an odd combination of styles.

A tall 'Spanish' Morion from the second half of the 16th century. This helmet was published in Antique Collecting Vol. 36, No. 8. (Private collection - Malta) Photo by M Bonnici

Front view of the helmet. This sturdy, heavy helmet was probably arquebuse proof. Photo by M Bonnici

Inside view of the same Morion. It retains traces of its original coarse cloth lining. Photo by M Bonnici

2.1.3 The Comb Morion

"Add to this, the pleasure of displaying on his official rides through the city—in shining contrast to the party-coloured gowns, half red, half tan, of the sheriffs and district officers—his fine military accoutrements, which you may admire to this day, sculptured on his tomb in the Valmont Abbey in Normandy, and his morion with all the bruises in it got at MontlhÊry...."

Notre Dame de Paris
Victor Marie Hugo
1831 AD

Some suggest that the Comb Morion is the latest form of Morion.[29] However, helmets having features of the helmet later to be called the 'Comb Morion' from 1535 and 1540[30] have been noted. At this point in the history of the evolution of this helmet the comb was low, the skull hemispherical, and the brim still horizontal. Towards the mid-16th century the brim was swept up, and as time went by the comb started to rise taller and taller.[31]

Comb Morions were lavishly embellished with embossing, etching, and gilding,[32] and it appears that the Comb Morion was always envisaged as being *'the most expensive morion.'* In fact, an example at the Louvre was made of gold and enameled. Comb Morions were frequently the hallmarks of town guards. The Comb Morions of certain town guards, such as those of the electors of Saxony, are much sought after by collectors. Comb Morions were produced in one or two pieces. The one-piece examples are considered to be the finest examples, because only armourers possessing an advanced skill were capable of moulding a singe sheet of steel into such a complex shape. The height of Comb Morions varies considerably. The height of Comb Morions varies from one centre of armour production to another, but the shortest Comb Morions tend to be around ten inches tall,[33] while the tallest in the range tend to be around thirteen inches tall.[34]

An Italian Comb Morion c. 1560. (Private collection - Malta) Photo by M Bonnici

Front view of the Comb Morion. It is made from a single sheet of metal. Photo by M Bonnici

This rare Comb Morion was made from a single sheet of metal—a very rare achievement. Photo by M Bonnici

2.1.4 The Burgonet

"When the King had seen this model, it restored him to cheerfulness. In the middle of the fountain I set a pedestal, projecting somewhat above the margin of the basin, and upon this a nude male figure, of the right proportion to the whole design, and of a very graceful form. In his right hand he raised a broken lance on high; his left hand rested on a scimitar; he was poised upon the left foot, the right being supported by a helmet of the richest imaginable workmanship. At each of the four angles of the fountain a figure was sitting, raised above the level of the base, and accompanied by many beautiful and appropriate emblems."

Benvenuto Cellini†
Autobiography
1562 AD
The Harvard Classics Translation 1909–14.

A French Burgonet from the second half of the 16th century. This helmet was published in Antique Collection Vol. 36 No. 8. (Private collection - Malta) Photo by M Bonnici

Profile of the same Burgonet (probably lot 701, Sotheby's Auction March 1995). The cheek pieces appear to be 19th century replacements. Photo by M Bonnici

Detail of the same Burgonet. This helmet appears to have been a 16th century relic that was restored in the 19th century. Photo by M Bonnici

The Burgonet was a semi-open faced helmet worn from the 1530s to the early seventeenth century.[35] It is often defined as being a metal helmet that consisted of a hemispherical bowl surmounted by a comb, having a neck skirt, a peak, and hinged cheek pieces.[36] In some examples the cheek pieces cover the face almost completely.[37] Some of the authorities on the development of Arms and Armour trace the origins of the Burgonet to a late sallet known as the 'burgundian sallet.'[38] However, these sallets lacked most of the burgonet's characteristics, and possibly the first burgonets evolved from the armet.[39] In fact, when burgonets were worn with beavors they were almost indistinguishable from armets.[40]

A very wide range of Burgonet patterns was used. These include the Cabasset/Burgonet that combined the pear stalk almond shaped skull of the Burgonet with the neck skirt and cheek-pieces of the Burgonet,[41] a Burgonet with three combs,[42] Closed Burgonets, and other patterns.

An early burgonet of armet construction c. 1540. *(Courtesy of Hermann Historica)*

Side view of the burgonet/armet. *(Courtesy of Hermann Historica)*

A 17th century closed burgonet, also known as a cuirassier's helmet.
(Courtesy of Hermann Historica)

A 17th century closed burgonet. *(Courtesy of Hermann Historica)*

A 16th century Burgonet with unusual acorn finial. *(Courtesy of Hermann Historica)*

The Burgonet was a helmet that was mainly used by the light cavalry,[43] but period artistic works prove that infantrymen of the officer class also wore it.[44] The Burgonet was used throughout most of the Elizabethan period, and it is not surprising that Shakespeare referred to this helmet in his *Henry VI*:

> "*I am resolv'd to bear a greater storm*
> Than any thou canst conjure up today
> And that I'll write upon thy burgonet
> *Might I but know thee thy household Badge*"[45]

Dating features on Burgonets include height and construction techniques. It appears that the skulls of early[46] burgonets were made out of one sheet of metal, while in later examples the skull was made in two halves. The presence of more than a single comb tends to indicate an earlier date, while it appears that burgonets with tall combs were produced towards the end of the 16th century. Simple and lower helmets appear to have been in production in the late 17th century.[47]

Burgonets were embellished with the same techniques used in Morions: etching,[48] guilding, and embossing.[49]

Battles Fought in Open Faced Helmets

"*Two pages were always by his side, one carrying his shield and helmet the other a pike....*"

Balbi Da Corregio
Diary of the Great Siege of Malta of 1565
1565 AD
Translation by Major Balbi 1971

Burgonets and Morions were extensively used in the battles of the second half of the 16th century and the first quarter of the seventeenth century. As light helmets they were ideal for sea warfare, and two of the most important 16th century battles happened to take place at sea.

Burgonets and morions appear on the heads of the combatants in some of the visual sources depicting the battle of Lepanto[50] of 1571 and the defeat of the Spanish Armada[51] in 1588.[52] Lepanto saw the humiliation of the Ottoman navy,[53] while the defeat of the Spanish Armada exorcised the threat of an invasion from the shores of England.[54] It is important to point out that morions and burgonets were not only used on the sea, but were also used on land. Open faced helmets were used in the decisive Great Siege of Malta of 1565, a battle that marked the end of the Ottoman myth of invincibility.

Unequivocal evidence of the use of the Morion in the Great Siege of Malta is contained in Matteo Perez D'Alleccio's fresco'*Assalto alla Posta Di Castiglia*. The contours of the 'Spanish' Morion can be clearly discerned on the defenders' heads on the lower centre of the frescoes. Some of the defenders higher up in the frescoes appear to be wearing similar helmets with curving brims, similar'to those of the peaked Morion. The printed version of D'Alleccio's painting *L'Assalto Alla Posta Di Castiglia* suggests that the Comb Morion may have been used as well. Great Siege period examples of all the three helmets are still stored in the Palace Armouries Museum today.[55]

The Figure of De Vallette in D'Alleccio's Assalto alla Posta Di Castiglia[56] suggests that the semi-open faced helmet[57] was also worn at the Great Siege of Malta. Judging from the numbers of existing specimens at the Palace Armouries Museum in Malta, it appears that open-faced and semi-open faced helmets were extremely popular with the storm troopers of Christendom, the Knights of Malta.

Burgonets and Morions were not only worn in the warm waters of the Meditteranean sea, but also in the Americas. Morions and burgonets appear to have been extensively used by the Spaniards in Colonial America.[58] Sixteenth century colored engravings, such as two examples in the *Bibliotheque de Service hydrographique de la Marine* in Paris, witness that the French used the Comb morion[59] and the 'Spanish morion' in the New World.[60]

2.4 Collecting

"*Each one took something from the heap—one a morion, another a rapier, a third a cross-hilted dagger.*"

Notre Dame de Paris
Victor Marie Hugo
1831 AD

Morions and Burgonets regularly appear for sale in the advanced auction houses and in the outlets of Arms and Armour dealers. Giving a price to these helmets is a very difficult task, for the price range depends heavily on the expertise of the seller and the purchaser's budget. The author recalls seeing a 'Spanish' Morion being offered for sale by an Italian dealer who had just acquired a large collection of arms and armour at a fraction of the price a similar helmet would sell on the international market.[61]

A considerable number of late 16th century plain[62] 'Spanish' Morions started to appear for sale in the early eighties in England, and the 'Spanish' Morions from this 'hoard' appear to have characterised the market for the past twenty years. Most of the helmets on the collectors' market appear to have been derived from what appears to have been a large group of quasi-identical helmets. Most of these helmets were probably arquebuse proof and tend to be rather heavy; they are also rather short when compared to other 'Spanish' morions.[63] In 1980 two examples were sold for as little as Stg85.[64] Two years later the same British auction house sold another two examples for Stg90 and Stg105, respectively.[65] In 1997 the same auction house sold another 6 quasi-identical morions for prices ranging from Stg130 to Stg180. Helmets from the same hoard[66] appeared at different dealers and other auction houses in England and the United States of America.[67] Some of the helmets from this hoard were struck with an armourer's mark on the rear brim. The mark consisted of a shield with pellets surmounted by a crown.

The author identified this mark as being a Spanish mark used in the 16th century.[68] It appears that the market has absorbed most of this group of helmets, and these plain morions no longer sell for such give-away prices. In June 2003 an American dealer was offering a helmet that probably came from this hoard at USD 2,500,[69] while a British dealer was offering an identical example for USD 1,920[70] and a blackened example with cheekpieces for USD 3,520.[71] The author believes that these

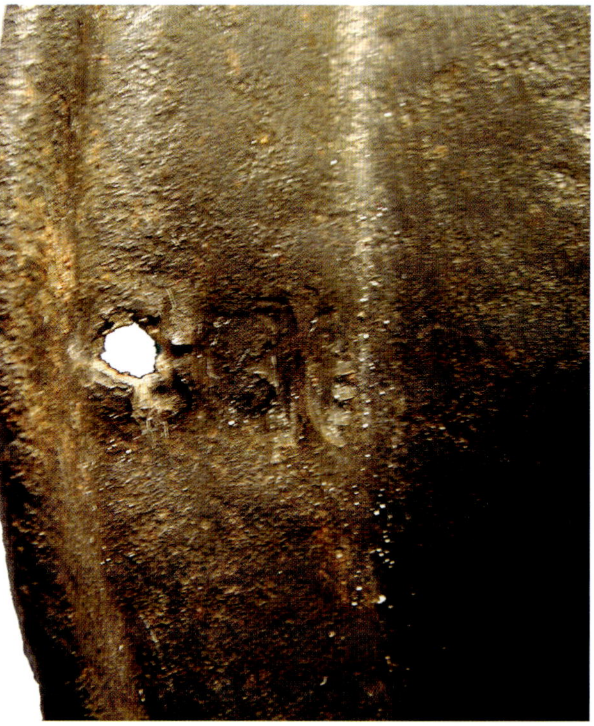

Close up view of the Armourer's mark on the Morion's brim. Photo by M Bonnici

excellent helmets may have some connection with the Spanish armada on account of their provenance, age, and original country of origin. In any case, these helmets certainly date from the reign of Elizabeth the first.

The rarest form of Morion is the Peaked Morion. Peaked morions rarely appear for sale. In June 2003 a British dealer was offering a very fine unetched example for USD 5,600.[72] Another example was offered for sale with 'buy it now facility' on ebay for the price of USD 2,190, but it remained unsold.[73] Comb Morions are more common, but owing to their elaborate shape and evocative appearance, even Comb Morions are very popular with collectors. Comb Morions were always expensive. A leading auctioneer valued a plain Comb Morion at Stg400-500 as early as 1995.[74] In 2003 a renowned British dealer was offering a plain Comb Morion for the price of USD 3,500,[75] but bargains are known to occur in flea markets and at non-specialised, unknowledgeable dealers. The author knows of Comb Morions that were sold for a fraction of their market price.

The prices for etched examples tend to be considerably higher. Etched Comb Morions are particularly in demand. A very finely etched late 16th century morion was offered for sale at Stg2,850 a few years back,[76] while another British dealer was offering a similar example at the price of USD 8,500 in June 2003.[77] The same dealer was offering a low Nuremberg town guard's Comb Morion stained with the distinctive 'black and white' pattern for the price of USD 4,450 and an etched 'Spanish' Morion at USD 2,250.[78]

The prices for good burgonets are generally slightly higher than the prices for morions. An excellent example of 'siege weight' was being offered for sale at USD 7,200 in June 2003.[79] Closed curassier's burgonets are much more expensive. An elaborate flemish example with fluted skull was being offered for sale at USD 16,500 in June 2003.[80] Simpler burgonets, especially the simple seventeenth century examples, are known to sell for much lower prices. These tend to sell in the region of USD 1,800 at auction.

Fake Morions and Burgonets abound. These may be classified in two categories; the modern replicas[81] and the old[82] replicas. The modern replicas would generally fool few on account of their bright 'white' colour[83] and the construction techniques used in the production of these helmets.[84] Modern replica morions and burgonets would, unlike some replica swords produced by modern armourers, fool very few people who observed and handled original items. Fine details, such as the construction of rivets, the signs of aging in the interior of the helmet, and the rolled edge are very difficult to replicate.

Old replicas are a different matter altogether. The Victorian era witnessed the production of some very well made embossed Burgonets and Comb Morions.[85] These are collectables and works of art in their own right that have an intrinsic financial value, but they are not *'the real thing'* and collectors should be careful. These replicas may tend to be rather tricky, as most of these replicas are over one hundred years old and have genuine signs of aging. Moreover, famous 19th century armourers, such as Gaetano Guidi of Florence,[86] created flawless helmets in terms of construction techniques.[87] Some of these fine helmets may be distinguished from their earlier counterparts by keeping in mind some basic guidelines. Victorian replicas were frequently made of cast iron, while earlier examples were never made of cast iron. Embossed Victorian replicas tend to be imitations of fine helmets in celebrated collections, such as the Louvres and the Wallace Collection. Examples of this caliber rarely, if ever, appear for sale. The quality of the embossing also rarely reaches the standards of the embossing on the original examples. However, it is admitted that Victorian replicas fooled and will continue to fool many experienced collectors and connoisseurs.

Notes:

[1] Blaire op.cit pg.150;
[2] Blaire op.cit pg.31;
[3] The War Illustrated 21st August '1915;
[4] The German Air Raid Defence, the *luftschutz* wore a helmet that was a kettle hat in every respect. Vide Attard Robert: The Kettle Hat (Chapel De Fer) published in the Armourer, The Militaria Magazine issue pp.22 and 23;
[5] Blaire op.cit pg. 138;
[6] Diary of the Great Siege of Malta of 1565 (Translation by Major H. Balbi);
[7] All quotations from Da Correggio's diary are taken from Major H Balbi's translation;
[8] Balbi op cit pg. 157;
[9] Balbi op cit pg. 154 ditto;
[10] Balbi op.cit pg. 137;
[11] These include Christie's September 1995 Sale Lot 78;
[12] It very hard to find Morions complete with cheek-pieces in museums as well as on the collector's market. For rare examples complete with cheek-pieces vide Muller Heinrich & Kunter Fritz: Europaishe Kelme figure 181 and Peter Finer Catalogue 1997 Lot 18;
[13] vide Laking op.cit nos. 31, 34 and 33;
[14] Traces of a canvas lining were noticed on a 'Spanish' Morion from the author's collection;
[15] vide infra;
[16] Christie's Arms & Armour September 1995 sale Lot 80;
[17] The Peaked Morion was a helmet which consisted of an Almond shaped skull finished with a tiny finial in the shape of a pear stalk. Its brim was swept upwards at the points of the peaks.. The Peaked Morion is the rarest type of Morion.
[18] The 'Spanish' Morion had an identical almond shaped skull to the one of the Peaked Morion. The difference from the Peaked Morion rested with the brim. In the Peaked Morion the brim was round and flat. The peaks of the crim were not swept up over the wearer's eyes and neck as in the Peaked Morion. The Spanish Morion was the most basic and cheapest helmet to produce on account of its simple construction.
[19] The Comb Morion was identical to the peaked Morion having the swept up peaks. The difference from the peaked Morion rested with the skull. The Comb Morion had a high comb surmounting the helmet skull. Roman Antiquarian recently advised the author that the purpose of the high comb was to serve as a hallmark for the Officers. The Comb Morion is used up till this day as the parade helmet of the Papal Guards.
[20] A very early Morion, which marks the development of the Kettle hat into the Morion, may still be picked out in the vast array of Morions in the Palace Armouries Museum[21] today, Vide Attard Robert: The Golden Age of Armour in the Palace armouries of the Knights of Saint John (Treasures of Malta). A similar helmet is found in the Royal Armouries in Spain and is reproduced in Laking's A Record of European Armour
[22] Laking fis. 93, 120and 188;
[23] Blaire fig.95;
[24] Blaire ip.cit pg. 153;
[25] vide infra;
[26] Muller fig. 101;
[27] Muller fig. 47;
[28] Cimarelli Aldo: Arms and Armour Figure 106;
[29] Cimarelli op.cit;
[30] Edge & Paddock pg 140;
[31] Muller fig. 82 and Blaire fig. 98 respectively;
[32] However it should be noted that the Armourers of Germany always tended to produce lower morions to their Italian counterparts. Blaire pg. 138;
[33] Muller figs. 83-87;
[34] Sotheby's July 1995 Arms & Armour Sale Lot 370;
[35] Comb Morion from the author's collection;
[36] op.cit Bull Stephen pp.92-93;
[37] Hayward Helena: The Connoisseur's Handbook of Antique Collecting pg. 51;
[38] op.cit Blaire pg. 137;
[39] op.cit Blaire pg. 136;
[40] So much so that an armet without its visor looks very similar to a burgonet, the only point of the difference is that the armet did not have a peak. Vide Hermann Historica April 2003 Sale Lot 325;
[41] Attard Robert: Death of an old Soldier published in Militaria international Magazine Volume 4 No.1;
[42] Tony Curtis: Lyle Price Guide Militaria Arms & armour 1993 pg. 187;
[43] Attard Robert: The Burgonet published in The Armourer Magazine Issue 26;
[44] op.cit Edge & Paddock pg. 140;
[45] Matteo Perez D'Alleccio: *Assalto Alla Posta Di Castiglia*;

[46] Henry VI Act V Sc1;
[47] Mid-16th century to third quarter of the 16th century;
[48] Examples noticed by the author. These include Sotheby's July 1995 Sale Lots 329 and 330 (early 17th century-both were around 17inches tall) and Sotheby's March 1995 Sale Lot 701 (16th century - around 11.5 inches - measured by the author);
[49] op.cit Muller figure Figures 50-53;
[50] op.cit Muller figure 58;
[51] Andrea Micheli: Battle of Lepanto (1539-1614);
[52] Zurbaran: Defence of Cadiz against the English (1634) (figures in the background);
[53] However it is interesting to note that Sir Francis Drake appears with a closed helmet by his side in some of his portraits. Closed helmets continued to appear in portraits as *objets d'art* up to the 18th century;
[54] Macintyre Donald: Famous Sea Battles pg.17;
[55] Macintyre op.cit pg. 29;
[56] Vide Guy Francis Laking and Attard Robert The Golden Age of Armour at the Palace Armouries of the Knights of Saint John (Treasures of Malta). Laking's Catalogue reports that there are around 400 helmets in the Palace armouries museum, most of these are morions;
[57] An inconsistency in D'Alleccio's works is observed here. Whereas in the Malta frescoes De Vallette appears to be wearing a burgonet in the Model paintings at Greenwhich the same figure appears to wear a Spanish Morion. D'Alleccio's visual evidence cannot be taken for gospel truth.
[58] The Burgonet
[59] Arms and Armour in Colonial America pg. 112 et seq;
[60] Graham Gerald S op.cit pg. 23;
[61] Graham Gerald S op.cit pg. 35;
[62] The helmet was offered to the author for the price of around USD 200;
[63] Unetched, acid cleaned, missing their liners, chinscales and almost invariably liner rivets etc;
[64] One of the shortest examples is in the author's collection. It is only around seven inches high;
[65] Wallis & Wallis Sale 258 Lots 1352 and 1353;
[66] Wallis & Wallis Sale 278 Lots 1642 and 1643;
[67] Apparently the provenance was an english manor;
[68] At very different prices. At times for as much as USD 800;
[69] Dudley S. Hawtrey Gyngell: Armourers Marks pg 79;
[70] LionGate arms and Armour;
[71] Blunderbuss antiques Lot 13873;
[72] Blunderbuss antiques Lot 17181;
[73] Blunderbuss Antiques Lot 17889;
[74] Item#2178175551;
[75] Sotheby's July 1995 Sale Lot 332;
[76] Fine Antique Arms unnumbered lot 6/7/2003 http://www.fineantiquearms.com/CatAHW.htm;
[77] Andrew Bottomley: Mail Order Catalogue No. 6 lot 554;
[78] Fine Antique Arms unnumbered lot 6/7/2003 http://www.fineantiquearms.com/CatAHW2.htm;
[79] Fine Antique Arms unnumbered lot 6/7/2003 http://www.fineantiquearms.com/CatAHW2.htm;
[80] Blunderbuss Antiques Lot 17892;
[81] Fine Antique Arms unnumbered lot 6/7/2003 http://www.fineantiquearms.com/CatAHW2.htm;
[82] Post c.1920;
[83] Pre.1920.
[84] Fakers can easily solve this telling mark by exposing their replicas to humidity or by staining them with greases and other oils.
[85] Comb morions tend to be constructed in 3 pieces or more (brim/skull/comb) and Peaked Morions in 2 pieces (brim and skull)—techniques that were never used in the 16th & 17th century. The 'pear stalk' on replicas tends to be of an odd shape, while the quality of the 'roping' tends to be poor.
[86] Andrew Bottomley: Mail Order Catalogue No. 6 lot 578;
[87] Cimarelli op.cit fig.15;
[88] Poorer quality replicas were produced in early twentieth century Germany. The embossing on these helmets tends to rather superficial;

3

The Zischagge and the Pikeman's Pot

From the English Civil War to the Thirty Years War

"So they consented and went, Mr Great-heart with his Sword, Helmet and Shield, and the rest with Spears and Staves."

John Bunyan†(1628–1688)
The Pilgrim's Progress.
The Harvard Classics. 1909–14

Claude Blaire, one of the greatest experts of armour, describes the 17th century as the period of the 'decline.'[1] The 15th century Armourer's ideal of 'beauty in function' was ignored by most of the armourers of the 17th century. Gone were the days of Anton Seusenhoffer, Piccinini, and other great artisans. Art had to give in to mass production. Finely chased and etched armour with the characteristic 'pots and mops' of the so-called Pisan school gave way to crudely forged, blackened armour without any frills. Aesthetic beauty was low in the priorities of the seventeenth century armourer. His priorities were of a purely functional character —durability, resistance to firearms, *arms blanche*, and possibility of mass production. This does not mean that the Armourers of the seventeenth century were not creative in their helmet designs. Around five different types of helmets were used in production

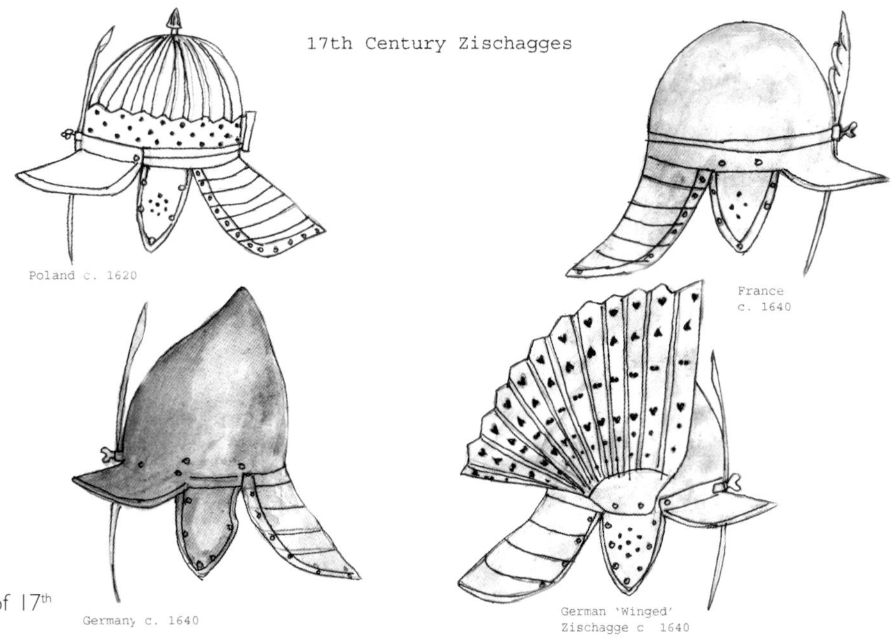

Artist's impression of a series of patterns of 17th century Zischagges.

in the seventeenth century. These were Zischagge (the Spider helmet), the Comb cap, the Pikeman Pot, and the Cavalier's helmet.

3.1 The Zischagge

The 17th century Zischagge was one of the most queer looking helmets to be mass-produced in Western Europe. It resembled the alien Kabuto.[2] The odd looking helmet possessed features that combined elements of various earlier helmets. The Zischagge's flamboyant neck flange was very similar to the lamellar nape protection of the 15th century Sallet.[3]

The Zischagge's movable twin cheek pieces and 'o' shaped finial were reminiscent of the Roman Troopers' Helmet[4] worn in Hadrian's time. In the Zischagge, these western European features were cemented with elements that were distinctly Islamic. The Zischagge's peak and, more notably, its movable nose guard were identical to those found on the Ottoman Onion Shaped Zucchetti[5] helmets of the 16th century.[6]

A 'Dutch' pattern Zischagge c. 1620. This helmet was published in Antique Collecting Vol 36 No. 8, and may have belonged to a Knight of Malta. (Private collection - Malta)† Photo by M Bonnici

Rear view of the Zischagge, with its characteristic lamellar lobster tail. Photo by M Bonnici

A flamboyant 'winged' Zischagge from the 17th Century. *(Courtesy of Hermann Historica)*

The First proper Zischagges[7] appeared in Eastern Europe[8] on the heads of the fierce Polish Cavalry Troopers. At times, the impressive helmets of the Polish cavalrymen were mounted with eye-catching[9] bellows shaped fans.[10]

In the Polish Zischagge, the exaggerated tall skull of the Ottoman pattern helmet was tailored down into a well-proportioned skullcap. A laminated neck flange constructed in the manner of a 16th century pauldron defended the shoulders, besides the wearer's head.[11] The neck flange was constructed out of a series of overlapping metal lames—the famous eye-catching[12] lobster-tail. This intelligent contrivance was a unique combination of mobility and protection. The neck protection was capable of extending and retracting with the movements of the wearer's head. Though this feature was known to the Gothic Sallet worn by the 15th century armoured Knights, it was welcomed as an innovation in the 17th century.[13] The movable cheek pieces of the Zischagge were punctured with Burgonet style 'salt shaker' vents that facilitated hearing. The wide leaf shaped nasal bar of the Ottoman Shishak was also reduced to a more slender steel bar. This offered the wearer of the Polish Zischagge a higher degree of visibility than his ottoman counterpart.

While this state of the art, 'high tech' helmet was being worn in the east, proving its effectiveness in the battlefield, the West still made use of helmets that were stultified in the Middle Ages. The Heavy Cavalry wore the totally impractical close helmet,[14] which placed the wearer's head in a badly ventilated steel cage. The Light Horse wore Burgonets.[15] The two helmets

Side view of the 'winged' Zischagge. *(Courtesy of Hermann Historica)*

above were heavy helmets, they were expensive to produce, and offered a degree of visibility that was significantly inferior to that of the Zischagge.[16] It is therefore not surprising that the west did not take long to adopt the eastern European contrivance.

By around 1620 Zischagges were being mass-produced in Western Europe. The first Western European Zischagges strongly emulated the designs of the Polish patterns.[17] The ribbed skull and, at times, even the brass ornamentation were retained, but slowly armour production centres started to produce distinctive patterns.

German and Dutch Armourers produced the successful design Blaire calls the 'Dutch pot.'[18] This was a simplified and cheaper version of the Polish Zischagge. The flamboyant neck flange, movable nasal guard, and cheek pieces were retained, but ornamentation was decreased to the very bare minimum.[19] The corrugated skull of the Polish helmet was done away with in the Dutch and German helmets. In lieu, in the Dutch and German patterns the skull was reinforced by a series of four radiating ribs. The run of the mill Western European Zischagges also missed another re-enforcing feature found in the Eastern European Zischagge. The Western European Armourers did away with the extensive ornamental riveting adopted by their Eastern European colleagues for the re-enforcement of the helmet. As the Italian scholar Cimarelli[20] points out, the extensive rivet patterns found on the eastern European helmets were not only decorative, but also served to give added resilience to the helmet. Thus, it may be concluded that the

An Eastern European Zischagge of high quality workmanship from the early 17th century. *(Courtesy of Hermann Historica)*

Zischagges produced in Western Europe were cheaper and inferior examples to their Polish and Hungarian Ancestors.

Yet, the Western European Armourer did not restrict his creativity to the Dutch pattern in his production of the Zischagge. England produced its own Zischagge during the Civil War. This helmet was destined to become synonymous with Cromwell's Roundheads and their fratricidal battles. Initially, Britain appears to have imported its Zischagges[21] from the continent, but as the demand for helmets increased, the Armourers of the London Armourers Company created their own helmet. The English Zischagge was based on the Dutch Pot. It retained the twin movable cheek pieces and the protruding design of the neck flange of the Continental helmet. Yet, it included innovations in the face guard and the construction of the skull piece, as well as the construction of the lobster tail.

In the Dutch Pot, the skull piece was fashioned out of a single piece of metal,[22] whereas the English Armourers fashioned their helmets out of two pieces of metal[23] firmly joined together to form a ridge.[24] The most apparent innovation of the English Zischagge revolved round the face guard. English Armourers must have found the protection offered by the sliding nasal guard of the Continental Zischagge inadequate, because they developed it into an improved alternative. The nasal guard was replaced with three fixed lames permanently attached to the helmet's peak and welded together to join at the wearer's chin. In the English Civil War pot the helmet's peak was no longer fixed permanently to the skull piece, but was joined to

Side view of the Eastern European Zischagge. *(Courtesy of Hermann Historica)*

the skull piece via pivots through which a 'closed helmet style' peak could be elevated and dropped.[25] This brainchild severed the Zischagge's Islamic heritage by doing away with the characteristic movable nasal guard.[26] The English alternative may have been more practical, as the face guard could be shut and swept over the helmet via the mobile peak. Another change in the construction of the English Zischagge involved the famous lobster tail. For practical reasons[27] English Armourers started to fake their lobster tails. Lobster tails were no longer made out of a number of lames, but were made out of a single sheet of metal incised to simulate a lobster tail. This change did away with one of the Zischagge's strongest assets.

The years of service of the Zischagge appear to have ended towards the late seventeenth century.[28] Yet, the Zischagge may have been retained in use in the 18th century, both as an extravagant parade casque,[29] as well as a reinforced siege helmet. Zischagges of very heavy, siege weight are known to exist,[30] but these cumbersome siege pieces were probably sappers' helmets or commanders' helmets, as suggested by Laking[31]—not true Zischagges. It is hard to imagine light cavalry troopers wearing these extremely heavy helmets, and the nature of the Zischagge was from its inception that of a light cavalry helmet.[32]

Whereas the early Zischagges produced in Eastern Europe were often decorated with lavish embellishment techniques, including incising,[33] blueing,[34] fluting, riveting,[35] and guilding, the Dutch pattern Zischagges and the English pattern Zischagges were rarely embellished. Some Zischagges appear

Italy c. 1660

England c.1642

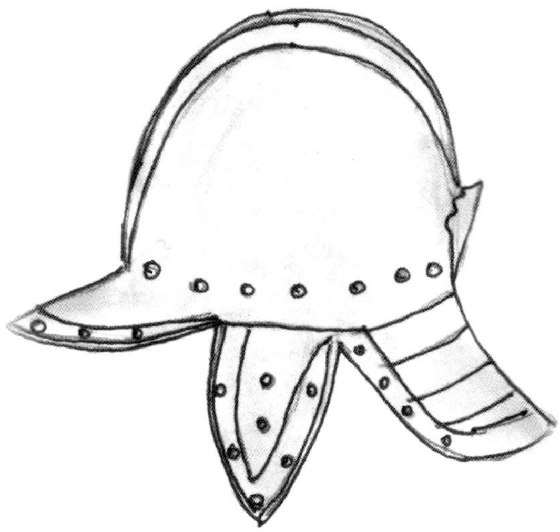

France c. 1680

Poland
c. 1620

RbA August 2003

Artist's impression of a series of patterns of 17th century Zischagges.

The Zischagge incorporated oriental characteristics, such as the nasal bar and the lamellar neck flange. Photo by M Bonnici

Close up of the Dutch pot's moveable nasal guard. Photo by M Bonnici

Internal view of the Dutch Pot. Photo by M Bonnici

Right: Close up of the armourer's mark on the helmet's peak. The author was unable to identify it, but it appears to be a German mark. Photo by M Bonnici

Below: Detail of the helmet's finial. The finial is reminiscent of some Roman helmets and was used for strapping. Photo by M Bonnici

An interesting relic Cromwellian cavalry trooper's helmet dating c. 1642-45. It appears to be an excavated helmet ex. Michael Dimbleby collection. (Private collection - Malta) Photo by M Bonnici

Despite its distressed state, the helmet retains faint traces of an armourer's mark stamped on the brim. Photo by M Bonnici

Back view of the same Cromwellian helmet. This relic helmet appears to be battle damaged. Photo by M Bonnici

This late sixteenth century backplate appears to have been excavated with the relic trooper's helmet. (Private collection - Malta) Photo by M Bonnici

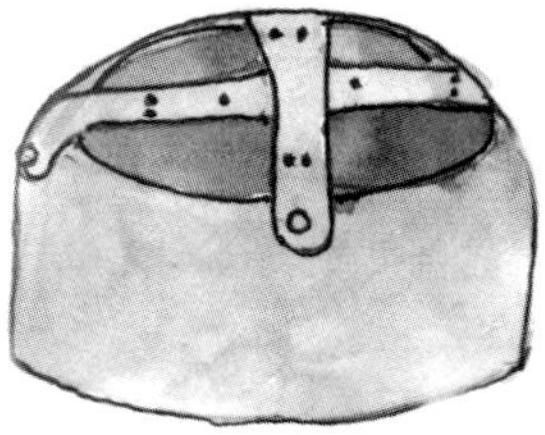

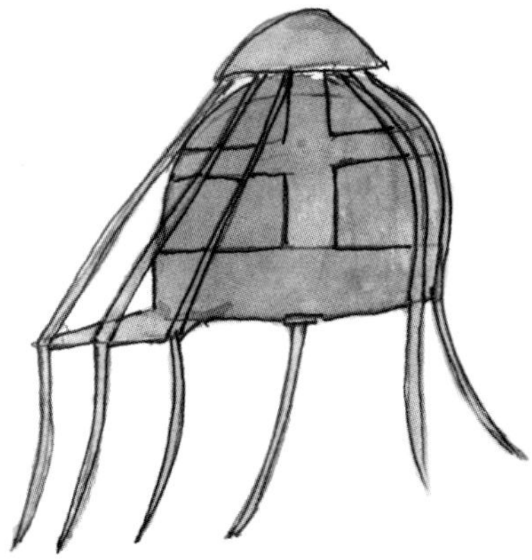

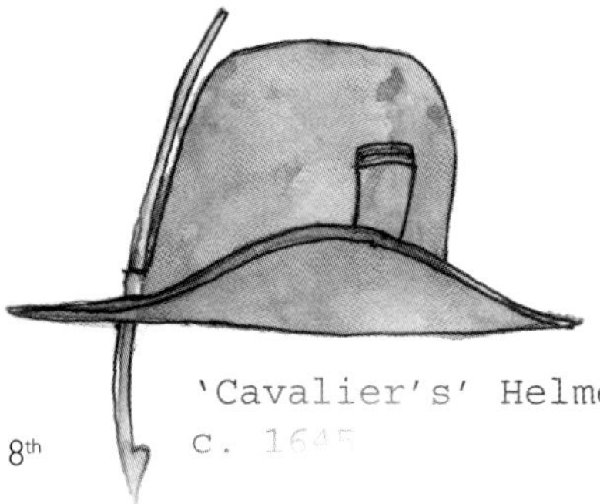

Artist's impression of helmets worn in the late 17th and early 18th centuries.

to have been blackened to protect them from rusting.[36] Others appear to have been painted with a thick film of paint.[37] Some rare English pattern pots made to measure for distinguished cavaliers were mounted with special high quality face guards. An example described by Blaire was fitted with a face guard fashioned in the shape of the Royal Coat of Arms.[38] Yet, these examples were special deluxe patterns which, unlike their coarse blackened counterparts, rarely appeared on the field.

Like other helmets of the time, Zischagges were fitted with special internal lining suspension systems.[39]These thick linings appear to have been made of a fine red cloth riveted onto thick leather straps that were riveted onto the helmet shell.[40] The whole interior of the helmet was lined except for the nasal bar. The entire neck flange, cheek pieces, and peak were covered with textile in such a way that the wearer's face did not touch the cold metal, and shock absorbed blows when the helmet shell was struck. The author never observed Zischagges complete with their linings on the collector's market.

It is important to note that rare variants of the Zischagges were in production in the 17th century. The first became known as 'the spider helmet' and is one of the most queer forms of metal headgear ever to be worn. The spider helmet consisted of a hemispherical bowl and peak identical in shape to the normal Zischagge. Unlike the normal Zischagge, the spider helmet was not fitted with a laminated neck guard and a nasal bar, but instead with a series of hinged strips of metal.[41] The second variation consisted of a metal version of the civilian hat worn at the time.[42] The President of the Court that tried Charles I, John Bradshaw, used a version of the latter helmet.[43] Both helmets are exceptionally rare, and the author did not notice examples of these on the collector's market. A fine specimen of a spider helmet was viewed at the *Musee d' l'Armee*, while a fine specimen of the cavalier's hat/helmet is kept at the Metropolitan Museum.

The Pikeman's Pot and the Comb Cap

"*Which a brown morion half-concealed,*
Light as a hunter's of the field;
And thus, with girdle round his waist,
Whereon the Banner-staff might rest
At need, he stood, advancing high
The glittering, floating Pageantry"

William Wordsworth
The White Doe of Rylstone
Or the Fate of the Nortons
(1770-1850)

In the first years of the 17th century the Armourers of the Low Countries and the Netherlands produced a variant of the 'Spanish' Morion. This helmet retained the characteristic almond shaped skull and narrow horizontal brim, but this helmet was made out of two halves and strengthened with a characteristic narrow comb '*with a step behind the apex*.'[44] Occasionally this helmet is referred to as 'the *Dutch Comb Cap*.'[45] The Arquebusiers in Jacques De Gheyn's famous *Wappenhandelinghe van Roers, Musqueten ende Spiese*—published in the first decade of—the 17th century—were portrayed wearing this type of helmet. Numerous examples of these helmets are kept in German, Belgian, and Netherlands Armouries,[46]†but these helmets seldom appear on the collector's market. Christie's offered a Dutch Comb Cap from the Howard M. Curtis collection for sale in 1995 and valued it at around USD 800.[47] These blackened helmets of a coarse appearance do not witness fine embellishments, such as those noticed on morions, and consequently sell at lower prices.

Cromwell's England produced its own version of the Dutch Comb Cap. The English comb-cap, known as a 'Pikeman's Pot,' had a very wide brim and a very short comb. On the earlier examples the brim drops down over the wearer's face, while on the post-1650 the brim is horizontal, and the helmet is generally of a smaller size.[48] The Pikeman's pot was, like the Dutch Comb Cap, worn with substantial lamellar cheek pieces. The 'Kettle-Hat' shape of the Pikeman's pot made it an ideal siege helmet, and commanders and siege engineers wore reinforced examples of siege weight.[49]

Collecting

When Guy Francis Laking wrote his *European Arms and Armour* earlier in the last century, 17th century armour, including the Zischagge, represented an accessible field to the average helmet collector. Unfortunately, this is not so any longer. When compared to the close helmet and the sallet the Zischagge is a common helmet, but when a Zischagge appears on the market it is faced by high demand. Eastern European Zischagges sell for thousands of pounds, while the cheaper Dutch and English examples are commanding high prices, as well. Even examples lacking a single cheek piece were sold in the region of Stg 500 a few years back. In 2003 a dealer who operates via the internet was offering a hungarian officer's Zischagge dating c.1630-1650 at USD 5,500.[50] A leading British dealer was offering for sale a complete 'dutch pot' produced circa 1630 at USD 4,450 and a complete Cromwellian armour complete with Zischagge at USD 10,000[51] in 2003. Another British dealer[52] was offering for sale two lobster tail helmets of the English civil war period at USD 4,560[53] and USD 4,240[54] in 2003. The same dealer was offering for sale a Polish lobster tail helmet c.1640 at USD 6,560[55] and a Victorian replica of a seventeenth century Polish cavalry helmet at USD 1,520.[56] Another dealer who has a sales portal on the net was offering a Zischagge at USD 2,800[57] in 2003.

For some strange reason the author cannot comprehend, the author failed to find a Pikeman's Pot being offered for sale in 2003. The Pikeman's Pot is by no means a rare helmet, but most pre-1900 militaria is becoming hard to find these days. The author would expect a good Pikeman's Pot to be offered for sale by dealers around USD 1,500.

Until recently the fake Zischagges that were on the market were no more than bizarre fakes that were made out of the wrong materials, and fashioned in patterns that did not exist in the 17th century. Very recently fake 'Hungarian' Zischagges of an eastern European inspiration appeared on the market. The helmets are flawless in terms of shape and construction techniques, but a close look at these helmets reveals that the patterns of these helmets do not tally with any known Zischagge. Some of the small components of these helmets are likewise oversized. The replica armourers' marks are also wrong in terms of size, and are in a characteristically modern†'type writer font.' Even the colour of the metal of these helmets appears to be too bright a white, but some of these forgeries were stained in grease; a forgery technique that is revealed by a characteristic smell. Victorian replicas of 17th century armour are easily recognised by their lightweight multi-plate construction and the inaccurate patterns used.

Battles worn in Zischagges and Pikemens' Pots

The Thirty Years War that started in Germany in 1618 was fought in Burgonets, Zischagges, and Morions. Cavalrymen wearing Zischagges appear in Jean Breitenfield's portrait of King Gustavus Adolphus depicting the '*Lion of the North*' at the battle of Breitenfield of 1631.[58] Another period illustration of the same battle illustrates Oberst Arnheum wearing an eastern European Zischagge.[59]

The same can be said of the English civil war, when an English variant of the Zischagges gave Cromwell's troops the name†'round heads.' Although the characteristic shape of the Spanish Morion may be identified in a period etching illustrating the raising of Charles the First's troops at Nottingham,[60]

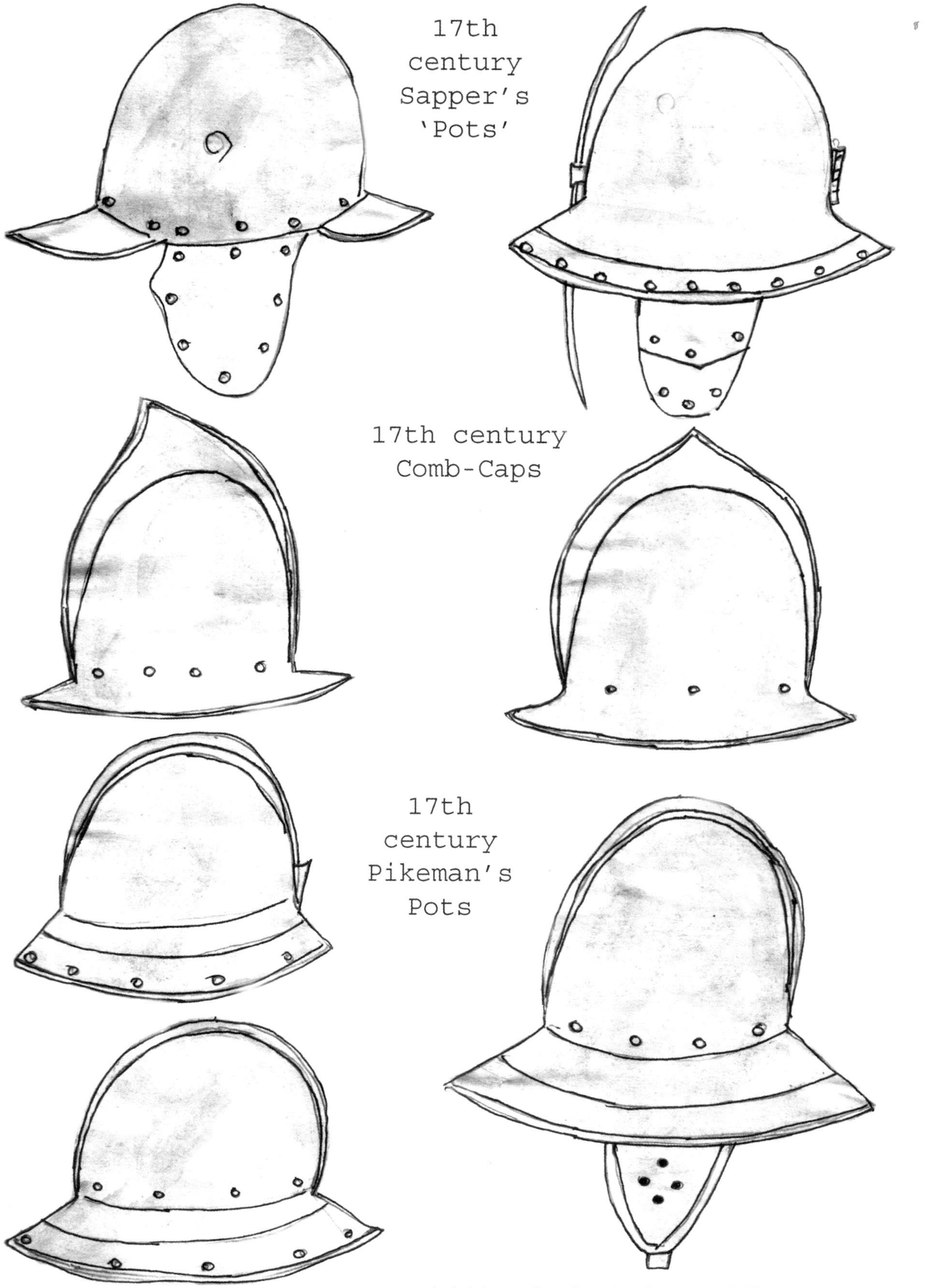

Artist's impression of a series of patterns of 17th century Pikeman's pots.

surviving specimens, including excavated specimens, confirm that English civil war battles, such as Naseby and Marstoon Moor, were also fought in three bar lobster tail helmets and Pikemens' Pots. Predictably, large quantities of these helmets are kept at the Royal Armouries in Leeds.

Notes:

[1] Blaire Claude: European Armour p.143;

[2] The Kabuto was the helmet worn by the Japanese Samurai.

[3] Blaire op.cit fig.110;

[4] An excellent portrayal of the Roman helmet and the Arms and Armour worn by the Roman Army in Hadrian's time is provided by Trajan's column in Rome;

[5] The Onion Shaped helmets worn by the Ottoman 16th century Sipahis were in their own turn an evolved pattern of the Kegelhelm of Ancient Greece. Vide Snodgrass Andrew: Greek Arms and Armour

[6] Boccia op.cit fig. 472;

[7] For an interesting array of photos illustrating different types of Zischagges and the evolution of the helmet consult the Ron Ruble enterprises Home Page.

The 17th century witnessed the occasional use of bizarre closed helmets, such as this Italian Savoyard c. 1600. *(Courtesy of Hermann Historica)*

Side view of the Savoyard. *(Courtesy of Hermann Historica)*

[8] Two beautiful eastern European Zischagges are found in the Palace Armouries Museum. Vide Laking Guy Francis: A Catalogue of the arms and Armour in the Armoury of the Knights of Saint John of Jerusalem fig.151 and Spiteri Stephen: The Palace Armoury A Study of a Military Storehouse of the Knights of the Order of saint John p.174;

[9] And totally impractical

[10] A rare example of this helmet from the Metropolitan Museum was illustrated in Bull Stephen: Arms and Armour p. 121;

[11] It is important to point out that the Zischagge was introduced in Europe when Armour was on its way out. Therefore the helmet had to act also as a shoulder protector.

[12] And aerodynamic

[13] Despite of its effectiveness, the neck flange of the 15th century Gothic Sallet seems to have been abandoned in the 16th century. The reason for this is unknown because it remains a superb invention, which may well be re-utilised by the makers of the modern kevlar combat helmets.

[14] The adoption of the Zischagge in Western Europe did not completely arrest the production of closed helmets. A Closed helmet which was as bizarre as the Zischagge was worn throughout the Zischagge's lifespan. This strange helmet was worn from around 1600-1660 and was called 'totenkompf' as it resembled a human skull. Vide Bull Stephen op.cit Cap. 4;

[15] Vide footnote, above

[16] In the Burgonet, the cheek pieces overlapped on the wearer's face blinding his side-ways view. In the Closed helmet visibility was limited to eye-slots punctured over the helmet's visor.

[17] For an interesting array of Eastern European Zischagges consult Lyle's Price Guide p.250;

[18] Blaire op.cit fig. 144;

[19] In some Polish examples cheek pieces were made out of more than a single lame, in the dutch pot cheek pieces were made out of a single lame. The radiated skull of the polish helmet was replaced with a round two piece skull having only four strenghtening ridges.

[20] Vide Cimarelli op.cit p. 32;

[21] Or at least the Dutch Pattern

[22] Excluding the finial holding the loop.

[23] Some English examples having skull pieces made out of a single sheet of metal are known to exist

[24] This method of construction echoed the construction of late Burgonets and Combed Morions, two other open-faced helmets worn in the early seventeenth century.

[25] A go-in-between version with fixed peak, two piece skull and sliding neck guard is known to exist Andrew Bottomley Catalogue fig.561;.

[26] Which was to be retained up to around 1870 in Persia and India, Vide Attard Robert: Islamic Helmets (The Armourer Magazine). Strangely enough even Burgonets with nasal guards were produced in the 16th century vide: Lyle's Price Guide p.237;

[27] Probably dicatated by the exigences of the blood thirsty civil war

[28] Vide Blaire Claude European Armour

[29] Vide 18th century painting by Antoine De Favray: Interior of Saint John' Co-Cathedral Malta (recently parts of this painting were enlarged and reprinted by Bank of Valletta) were a standard bearer is depicted wearing what is probably a Zischagges without cheekpieces.

[30] Vide Andrew Bottomley Catalogue fig. 563;

[31] Vide Guy Francis Laking: Arms and Armour at the Palace Armouries of the Knights of Saint John of Jerusalem. It should be pointed out that the examples described by Laking are an unusual combination between the Burgonet and the Zischagge.

[32] Foot soldiers did not wear Zischagges but wore the Pikeman's Pot while the Heavy Cavalry retained the anachronistic close helmets of the previous century. Vesey Norman: Arms and Armour

[33] Vide Galea Michael: : The Palace and the Palace Armouries of the Knights of Saint John

[34] Vide helmet illustrated full page in Bull op.cit p.111. It is not known whether this helmet is actually eatern European. Bull suggests that it may be French.

[35] Vide Zischagges illustrated Cimarelli op.cit;

[36] Vide the two helmets in the panopolies illustrated in Bull Stephen's Arms and Armour.

[37] A Dutch Pot in the Author's Collection which appears to have been cleaned some years back retains minor traces of thick graphite paint.

[38] Vide Blaire op.cit fig.55. It is interesting to note that Charles 1st also appears to have worn a dutch pot as illustrated in Cimarelli Aldo: Arms and Armour. His colleague Louis the XIV appears to have opted for a Zischagge as well. Vide Vesey Norman: Arms and Armour

[39] Cimarelli Aldo's Arms and Armour includes illustrations of rare Zischagges retaining their original linings.

[40] And held to the helmet via square shaped hand forged washers.

[41] Cameran Stone George: A Glossary of the Construction, Decoration and Use of the Arms and Armour in All Countries and In All Times fig. 749;

[42] Stone fig. 389;

[43] Snellgrove L.E: The Early Modern Age p.165;

[44] Blaire op.cit p.153;

[45] Christies September Sale 1995;

[46] Muller op.cit figs. 107 and 108;

[47] Lot 76 Christie's September 1995 sale;

[48] Blaire op.cit p.154;

[49] Muller op.cit p.177;

[50] www.cullodenantiques/Zischagge.html;

[51]http://www.fineantiquearms.com;

[52] Blunderbuss antiques;

[53] Lot 17894;

[54] Lot 17041;

[55] Lot 17896;

[56] Lot 11619;

[57] http://www.liongate-armsandarmour.com;

[58] Macdonald op.cit p.27;

[59] Snellgrove op.cit p.138;

[60] Snellgrove op.cit p.150;

4

The 18th Century
The Decline of the Metal Helmet from Blenheim to Saratoga

By the mid-17th century the popularity of armour was in decline. Advancements made in the production of firearms signalled the death knell of the metal helmet. The success of the Pikeman's pot was short lived, and the cavalry disposed of their Zischagges altogether. The 18th century witnessed the total eclipse of the metal helmet; the infantry was equipped with soft and semi-soft headgear that offered little protection. Almost all forms of armour were considered to be anachronistic, and the use of armour was reduced to the very bare minimum.[1] The metal helmet was destined to return to the infantry around two hundred years later. In the mid-18th century the cavalry started to make a half-hearted use of leather helmets[2] fitted with substantial metal plates and fittings. In the first years of the nineteenth century these helmets developed in the neo-classical helmets so popular in the Napoleonic wars.

An exceptionally rare type of fatigue cap, a Phyrigan cap from the French Revolution.

4.1 The Mitre Helmet and the Bearskin Grenadier Cap ('tall caps')

4.1.1 Military Exigencies

"Mr. Stubble, as may be supposed from his size and slenderness, was of the Light Bobs. Ensign Spooney, on the contrary was a tall youth, and belonged to (Captain Dobbin's) the Grenadier Company, and he tried on a new bearskin cap, under which he looked savage beyond his years."

William Makepeace Thackeray
Vanity Fair
19th century

Contrary to popular belief, the hand grenade is not a weapon that was created in the 20th century. Surviving examples prove that earthenware hand grenades were used in the crusades. Grenades were extensively used in some 16th century sieges.[3]

The wide-brimmed headgear issued to the infantry in the 18th century was probably inadequate for the use of this effective incendiary weapon, and some believe that the tall Mitre helmet and massive Bearskin Grenadier cap were created to solve this problem. These tall caps were tight around the skull and could

Artist's impression of Mitre helmets from the 18th and 19th centuries.

A Prussian stamped brass mitre helmet of Infantry Regiment 37 c. 1740. *(Courtesy of Hermann Historica)*

4.1.3.3 France

French bombardiers wore similar mitre helmets to those worn by their Prussian counterparts at the time of Louis XV,[23] but as time went by the metal element in the tall cap was gradually decreased in favour of the fur component. In the second half of the 18th century French grenadiers wore tall bearskin caps fitted with substantial brass or gilt leather helmet plates embossed with the bourbon coat of arms.[24] The size of these helmet plates shrunk as time went by. The logo on the helmet plate also varied from one regime to another. The bourbon coat of arms was replaced with the legend *Vivre libre ou mourir* after the French revolution.

An exaggeratedly large bearskin cap was introduced in the French Army in October 1801.[25] It was the black Bearskin Cap we tend to associate with Napoleon's bearded grenadiers, the legendary *Bonnet d'Ourson*. The Model 1801 bearskin was thirteen inches high and fitted with a gilt helmet plate embossed with a grenade motif, and a scarlet cloth backing braided with a gold cross. In 1808 the gold cross motif was replaced with a grenade motif. The bearskin was embellished with cap lines and plumes; the colours of these accessories varied from regiment to regiment. Napoleon's grenadiers used to employ their heavy bearskins only on parade and as combat helmets, while the grenadiers used to wear Bicorn Caps on undress

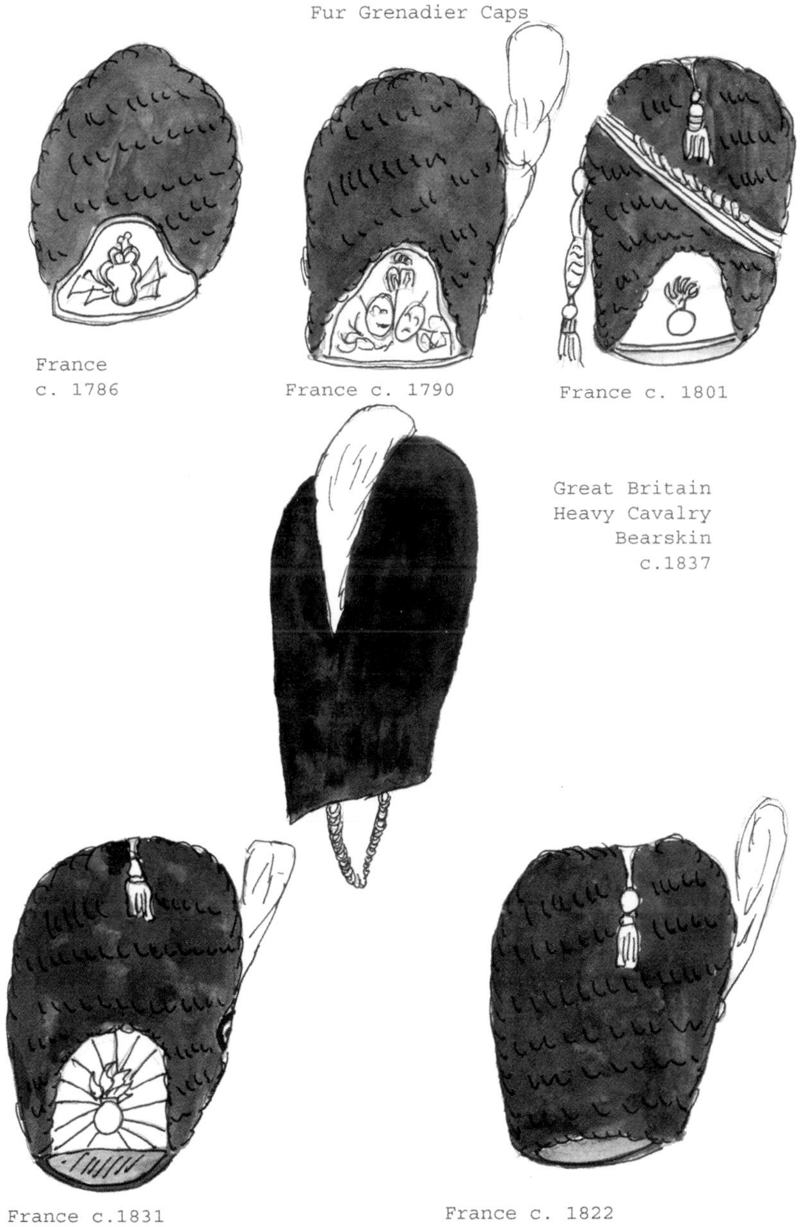

The evolution of the *Bonnet d'ourson* from the 18th to the 19th century.

duties. Napoleon's grenadiers were not the only Napoleonic troops to wear the bearskin. Even the *Chasseur a' Pied* of the Imperial Guard, the *Carabiniers a Cheval*, the *Grenadiers de la Garde Imperiale*,[26] the *Artilleurs A Pied*, the *Gendarmes de la Garde*, the *Sapeurs*, and the *Grenadiers a' Cheval*[27] used to wear bearskins.[28] *Artilleurs A Pied* and the *Gendarmes de la Garde* used to wear a special pattern bearskin fitted with a leather peak.

Napoleon's decline and fall did not signal the death knell of the *Bonnet d'Ourson*. The newly formed *Garde Royal* included six Grenadier regiments equipped with bearskins[29] fitted with a helmet plate embossed with the bourbon coat of arms. The *fleur de lys* of the bourbons was replaced with a grenade motif in 1830 with the fall of Charles X.

Needless to say, the bearskin was retained in use at the time of the second Empire when it was fitted with a helmet plate featuring a Napoleonic eagle.[30] The bearskin fell in disuse in France with the fall of the second empire.

An exceptionally rare relic Grenadier Cap of a Grenadier of the French *Garde Nationale*, complete with its scarlet *rabattre* (cap chords) c. 1830. (Private collection Malta) Photo by M Bonnici

Close up view of the Grenadier Cap's plate. Photo by M Bonnici

Inside view of the Grenadier Cap. Photo by M Bonnici

4.1.3.4 Great Britain

The Grenadier regiments of Britain wore various forms of Grenadier caps. English grenadiers wore fur caps as early as the late 17th century. By the early eighteenth century the stiffened flat mitre cap replaced the bearskin.[31] The English mitre caps were entirely made of coloured felt. No metal plates were used. The caps were marginally strengthened by intricate patterns of embroidery.[32] Colonel Rankin describes this odd and expensive manner of construction as 'a British peculiarity,'[33] but it appears that some of the American grenadier companies that fought the American war of Independence wore 'home made' copies of these helmets.[34] In England the examples worn by officers were made of buff velvet instead of felt, and the embroidery was made in gold and silver thread.[35] In 1768 the British Austrian style bearskin caps substituted the stiffened felt caps. The Model 1768 caps consisted of bearskins sewn into a mitre shape and strengthened by large metal helmet plates.[36] Some British cavalry regiments, such as the Scot's greys, wore similar helmets fitted with leather peaks.

The M1768 cap fell into disuse in a short time and disappeared in the beginning of the 19th century, when it was replaced with the Shako. The success of the French grenadiers appears to have inspired British military tailors as much as they inspired Spanish, Danish, Swedish, and Italian tailors, because after the Napoleonic wars the British army started to wear large bearskins[37] identical in shape to the bearskins previously worn by the French. The First Foot Guards were equipped with bearskins right after the battle of Waterloo.[38] The bearskin cap was also issued to the Life Guards in 1821.[39] Similar bearskin caps were worn in Sweden, Denmark,[40] and Sardinia. The British Guard regiments still wear identical bearskins today.

A British Grenadier Guard's Bearskin c. 1970,. The bodies of these helmets are frequently employed in the construction of forgery and replica Napoleonic bearskin caps. Photo by M Bonnici

4.1.3.5 Imperial Austria

The Imperial Austro-Hungarian Grenadier regiments fought Napoleon's Old Guard with Grenadier Caps similar to those worn by their French foes on their heads. These tall fur caps started to be used in Imperial Austria in the last years of the 18th century, and remained in use up to the first years of the 20th century. The shape of this headgear evolved very little in its history. In 1799 the Austrian grenadier cap consisted of a tall fur cap with a large brass plate. In this early phase of its history the Austrian fur cap was worn without peaks and brims,[41] but it was worn with a substantial coloured cloth bag.[42] In the early 19th century the fur caps were fitted with a leather brim, a rare peak, and two side peaks,[43] but it appears that by 1830 Austrian grenadiers reverted to the use of a fur cap peak without brim.[44] Around thirty years later the fur cap was fitted with a brim once more.[45]

Front view of an exceptionally rare Austrian Grenadier's Cap from the second half of the 19th century. (Private collection Malta) Photo by M Bonnici

Close up of the Grenadier Cap's helmet plate. It is embossed with the Hapsburg eagle. It appears to have been mounted with battle honours. Photo by M Bonnici

The silk liner of the 19th century cap. It bears the label of the famous firm Leopold Verch (which later became a famous film costume company). It also bears the stamp of the IR XXXVI Regiment of Klein. Photo by M Bonnici

Close up of the helmet's reverse. Note the moth and age marks. Photo by M Bonnici

View of the reverse side of the Grenadier Cap. It is heavily mothed, but is miracoulsly complete. Photo by M Bonnici

4.2 The *Casquette* and the Tricorne

"...the King in his Windsor Uniform with the star of the Order of the Garter on his chest and wearing a hat with a gold button and loop, surmounted by a black cockade, which marked him 'out conspicuously from the rest of the company.'"

Christopher Hibbert
George III (1738-1820)

The tall cap was a mark of distinction used by chosen elite regiments. The rest of the infantry wore the Tricorne hat. The Tricorne hat was extremely popular in the 18th century, and was also used by the navy and the cavalry. The latter also made use of a more elaborate piece of headgear, a leather helmet derived from the Burgonet, the *Casquette*.

The Tricorne

The Tricorne hat appeared in the late 17th century[46] and was conceived as a civilian hat,[47] rather than a purely military hat. In fact, the Tricorne was extensively used as a civilian hat in the 18th century. The shape of the Tricorne developed from the non-standardised, simple wide brimmed '*sombrero*' style hats worn in the 17th century.

In the 17th century the right side brim of the wide-brimmed hat started to be turned, or 'cocked' inward.[48] At the time of the English civil war the hat was worn with a single cocked flap, and a metal variant of this hat was in production in Cromwell's days. Later on, the two lateral flaps of the brim started to be folded inwards[49] in the 'Dutch style.' A few years later even the back flap of this hat was folded inwards. Thus, the hat

One of the rarest and most representative forms of 18th century headgear, an English Officer's Tricorne c. 1770. *(Courtesy of Hermann Historica)*

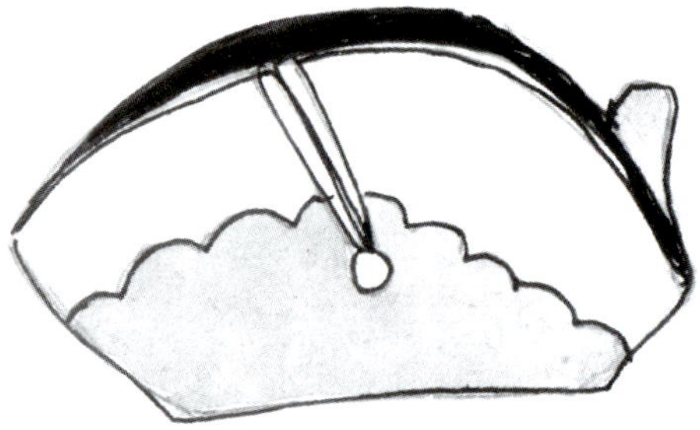

Artist's impression of 18th century Tricornes.

developed the distinctive three-pointed shape that gave the Tricorne hat its name. This manner of cocking the hat may have served a useful purpos; the three little spouts served as a conduit to clear water from the hat's fragile surface.

Tricorne hats were made of heavy black felt reinforced with gold, silver braiding, or a cloth tape woven around the brim. Occasionally, the brim was trimmed with animal furs or feathers. A colourful cockade was attached to the hat through a metal button, and a pair of 'V'-shaped ribbons were secured to the button to strengthen the front part of the hat. The cockade and the button served as a means of identification. Needless to say, such a soft hat could not offer serious protection to its wearer, so the Tricorne hat was occasionally worn with an iron cross piece that protected the exposed skull from sword cuts.[50]

4.2.2 The *Casquette*

Cavalry regiments extensively used the Tricorne in the 18th century. An Ordinance enacted on June 1, 1750, records that the Tricorne was the hat worn by the *cavaliers* of France.[51] However, the Tricorne hat was inadequate as a military headgear, and the need was felt to equip the cavalry with a leather and metal 'jockey' helmet that offered more protection to its wearer.

This form of headgear appeared in the late 17th century[52] and was extensively used in the battles of the eighteenth century. This helmet became synonymous with Banastre Tarleton, a British cavalryman who fought in the American war of independence. Leather crested

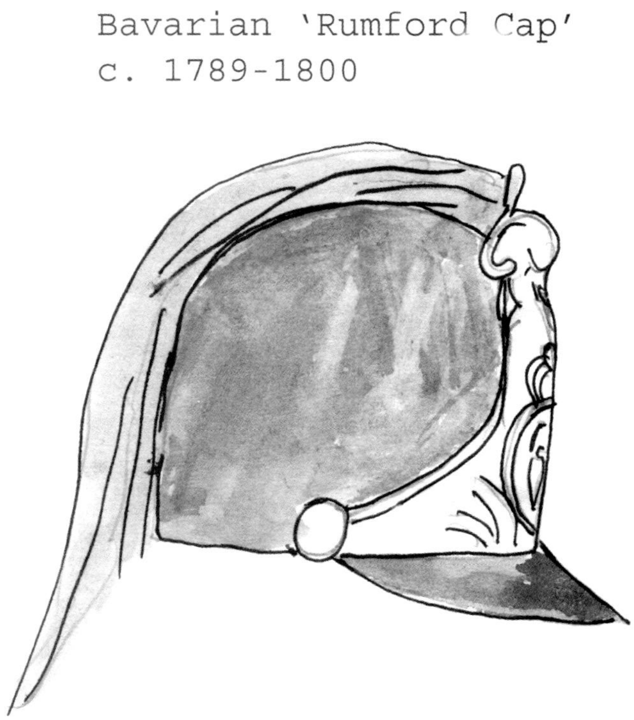

Artist's impression of an 18th century Bavarian raupenhelm.

Casquettes of burgonet form were worn by most of the dragoon regiments of the 18th century. Most of the patterns worn by the colonists in the American war of independence appear not to have been standardised.[53] Some examples appear to have been 'homemade.'

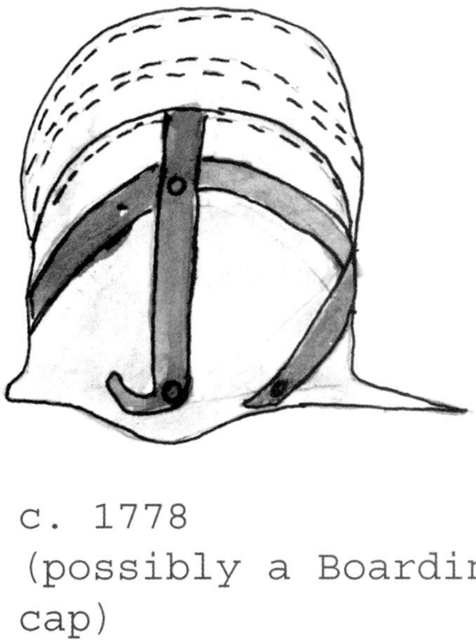

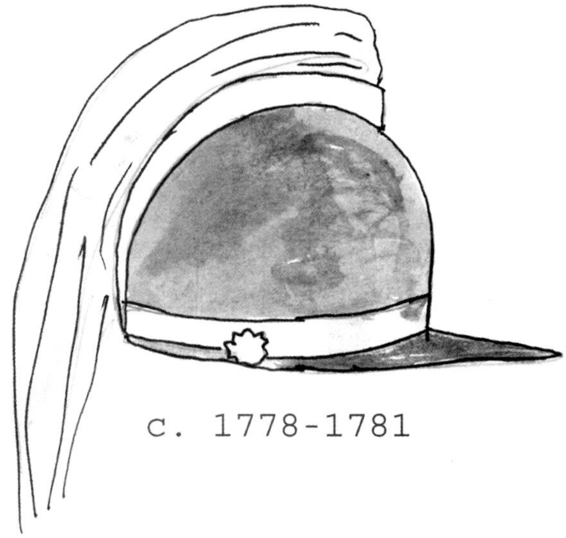

Artist's impression of two 18th century U.S. casquettes.

4.3 The *Haiduk* and the Fatigue Cap

"The fatigue cap is the same colour of the uniform, dark blue trimmed with scarlet facings...."

M. Malibran
Military Regulations October 1786[65]

The cavalry regiments of Eastern Europe were highly respected on the battlefield, and their uniforms and traditions were emulated in Western Europe. In the 18th century Hungarian hussars wore a cylindrical hat that resembled a tall fez wrapped with a long band of cloth. The loose cloth strap was left to fly in the wind, and this was the reason why the Germans called this hat '*Flugelmutze*,' meaning 'wing cap.'[66] This hat started to be worn in Europe as early as 1743 when it was known as a 'Hungarian style headgear,'[67] or simply '*Mirliton*' or '*Shakelhue*.'[68] The cavalrymen of France, England, Russia, some German states, and Sweden wore the Mirliton. In the late 18th century this hat developed into the Shako and fell in disuse. However, it made a short re-appearance in the Second World War when Mussolini equipped his native cavalry with a scarlet tall fez that had some common denominators with the Mirlitons worn in the 18th century.

Flamboyant and expensive headgear, such as the Haiduk, Tarleton, Mitre helmet, and tall bearskin cap must have been tremendously effective on the battlefield. Mitre helmets and bearskins magnified the size of their wearers, giving them the appearance of titanic giants, but these hats were very heavy and uncomfortable. Softer and lighter hats were worn by the troops for undress wear. These light caps became known as fatigue caps, and versions of these hats are still worn today.[69]

Various forms of fatigue caps were worn in the 18th century. The fatigue cap known as the *Zipfelmutze* was particularly popular on the European continent. It was worn in Austria,[70] and in France it became known as '*Bonnet de Police*,' '*Bonnet a La Dragonne*,' or '*Bonnet a flamme*.' The *Zipfelmutze* consisted of a colourful soft pointed felt cap of a triangular shape with a '*tail*' that ended with a tassel. The tall point was folded sideways. The *Zipfelmutze* was probably derived from military headgear worn in the Ottoman Empire in the sixteenth century. The French dragoons formally adopted it in 1740. It was issued to other branches of the cavalry, the carabineers, and the light cavalry in 1786. It was later issued to the cuirassiers. A minor change to the shape of the *Zipfelmutze* was made at the time of the First Empire, when the part of the hat worn round the crown of the head was stiffened in the shape of the modern overseas cap. The shape of the *Zipfelmutze* became more similar to that of the overseas with the changes made in 1822 and 1837, when the size of the hat and tail was further reduced.

Artist's impression of an 18th century German Flugelmutze of the Von Ruesch Hussar Regiment.

Other types of fatigue caps were worn in the eighteenth century. A variety worn by most of the armies of Europe consisted of a tent shape hat with a cloth peak and brim that could be folded outwards.

Battles Fought in Tall Caps, Haiduks, Tricorne hats, and Tarletons

"*...with padded shoulders and tall helmets or bearskins they looked supermen....*"

John Macdonald
Great Battlefields of The World

The elite regiments who fought the wars of the second half of the 18th century and the Napoleonic wars wore tall caps. The shape of the stiff felt mitre shaped hat can be discerned in the period illustrations of the battles of the American War of Independence,[71] and on the heads of the British troops enrolled to quash the Scottish rebels at the battle of Culloden.[72] Similar helmets appear on the heads of Frederick the Great's Grenadiers in period paintings.

A *Zipfelmutze* (bonnet de police) model 1786 of the Cuirassiers c. 1791-1815. *(Courtesy of J. P. Morange)*

View of the *Zipfelmutze's* liner. *(Courtesy of J.P. Morange)*

Side view of the *Zipfelmutze. (Courtesy of J. P. Morange)*

Side view of the *Zipfelmutze. (Courtesy of J. P. Morange)*

In the Napoleonic Wars the tall bearskin cap became synonymous with two of the finest grenadier regiments of the period; the French grenadiers and the Austrian Grenadiers. The bearskin was a mark of distinction in the Imperial French Army. The expensive hat was only worn in battle or on parade. Captain Coignet describes how the grenadiers wore their bearskin at the very last minute, and they did not have time to embellish their bearskins with their regalia at the battle of Wagram.[73] It was also a grenadier wearing a bearskin ferociously wielding an axe that stormed into Hougomont, while *Grenadiers a Chevals* wearing bearskins charged the squares at the battle of Waterloo.[74] Bearskins must have also turned out to be very useful in the Russian campaign.[75] The headgear of the Austrian Grenadiers greatly intimidated some of the French troops at the Battle of Aspern Esslingen, when some French soldiers erroneously believed the Austrian grenadiers to be Austria's equivalent of the Emperor's feared *Vieille Guard*.[76]

Tricorn hats were worn in most, if not all of the important battles of the eighteenth century. They were worn by Frederick the Great's infantry[77] and cavalry alike.[78] The tapestries at Blenheim Palace depicting the battle of Malplaquet[79] portray Marlborough and his officers wearing Tricorns. Similar headgear appears in depictions of the battle of Blenheim[80] and other major 18th century battles. Tarletons and Haiduks appear

Rear view of the *Zipfelmutze*. *(Courtesy of J. P. Morange)*

in various 18th century portraits, but it is very difficult to identify these types of headgear in period battle illustrations. One of the few exceptions is an 18th century watercolour illustrating scenes from the Turkish wars. Tarletons appear in most of the period portraits of Banastre Tarleton, as well as other officers of the American War of Independence. These include the portrait of Captain Thomas Youngs Seymour at the Connecticut Society. Visual evidence attesting the use of Haiduks in the eighteenth century is found in the detailed portraits of Chevalier de Grassin at the *Musee de l'Armee* and a portait of an Officer of the Hussar Regiments of Camborant at at the *Musee' De l' Emperi.*

Collecting

"One of the most interesting items of military headgear is the grenadier cap."

Col. Robert Rankin
The Official 1983 Price Guide to Military Collectables

Eighteenth century headgear is exceptionally rare—rarer than most 16th and 17th century forms of headgear. Most of the military headgear worn in the 18th century was constructed from delicate materials, such as felt, cloth, and leather. When rare specimens of 18th century headgear appear for sale they tend to sell for very high prices. Wallis and Wallis auctioned a British Grenadiers 1768 Fur Cap for around USD 8,000 in 1993. Two years later the same auction house offered a unique opportunity to acquire an unidentified 18th century *Zipfelmutze,* but the offer was not taken on the bidding floor.[81]

Recently Sotheby's offered a splendid collection of dragoon helmets from the American War of Independence for sale.[82] All the extremely rare pieces came from the William Guthman Collection, an extraordinary collection that included very hard-to-find militaria.[83] Museums and prestigious historical societies competed for some of the items. A colourful Tarleton was purchased by a Connecticut collector for USD 45,600, while the Connecticut Historical Society acquired another fine dragoon helmet for USD 44,000.[84] British Tarletons from the Napoleonic were always popular with collectors, and the prices of these rare helmets reflect this popularity. Blunderbuss Antiques offered a cavalry tarleton of the King's German Legion c.1810-1815 in a very good clean condition for USD 10,400,[85] and a Tarleton of the Indian Army from 1800 for USD 8,800.[86] The author recalls noticing a relic Tarleton stripped of most of its fittings and its characteristic bearskin crest being sold for around USD 1,500 in an internet auction.

Haiduks are much harder to find than *Casquettes* and seldom appear for sale. A French dealer is presently selling a late 18th century Haiduk in very good condition in the region of USD 9,000.[87]

Most of the mitre helmets that appear for sale are Model 1894 Prussian parade Mitre helmets. They are in very high demand and tend to sell for four digit figures. A fine example was recently offered for sale in an internet auction with 'buy it now' facility for USD 3,500, but this rare helmet remained unsold.[88] Mohawk Arms recently sold a very fine example complete with its original oil cloth cover for USD 3,800. The Old Guard, a dealer who specialises in Napoleonic militaria, offered a French bearskin plate c. 1809-1815 at USD 5,500 and a helmet plate of the Pavlovski Grenadier regiment at USD 3,000.

Twenty years ago, Colonel Rankin wrote of various 18th century grenadier caps being sold for prices ranging from USD 4,000 to USD 9,000, but very few 18th century mitre helmets appeared for sale in the last few years. An exceptionally fine and extremely rare collection of 18th century headgear was offered for sale by Hermann Historica[89] in September 2003. This auctioneer valued a Russian mitre helmet (c. 1802-1807) at around USD 5,000[90] and a mitre plate of the Prussian grenadier regiment at around USD 6,000.[91]†A pair of complete 18th century caps were valued at around USD 20,000 each,[92] and an exceptional 18th century bombardier's cap (possibly the only example 'not in the possession of a museum') at around USD 20,000.[93] Replica 18th century mitre-helmet plates are known to be in production, and although the casting of the 'mirror' plates lack the fine details of the original examples, only a very close inspection can tell the wheat from the chaff. It is also important to keep in mind that replica Prussian 18th century *Grenadiermutze des Infanterie-regimentes v. Wunsch* were being produced in the late nineteenth century, and that these helmets are almost impeccable in terms of construction techniques.

Tricornes are a veritable minefield and even the greatest connoisseurs would hesitate before authenticating this unobtainable form of headgear. The author noticed knowledgeable auctioneers showing signs of insecurity and using phrases such as '*possibly 18th century*' or '*probably original*' in relation to Tricornes. Flawless 18th century Tricornes were worn and produced for the British Chelsea Hospital as late as 1950, and these hats are a bombshell in the wrong hands. Even some examples that are presently being hand sown for re-enactments may deceive some collectors. Original Tricornes seldom appear on the collector's market. The author noticed a good Tricorne being offered in 1995 when Kent sales offered a mid-18th century Tricorne hat with a button of the Royal provincials valued at STG 250-400.[94] Recently Hermann Historica valued a superb English Tricorne c. 1770 at around USD 2,500. The same concerns that were expressed in relation to Tricornes may be expressed in relation to Victorian bearskins; the construction techniques used in the production of British guardsmen's bearskins in the 'seventies are very similar to those used in the 19th century.

Zipfelmutzes are very hard to find, and the author only recalls two 1840 Belgian *Zipfelmutzes* being offered for sale in the past year or so. They were being offered for sale for a pittance, around USD 100. The *Zipeflmutze's* similarity to modern tent pole caps tends to decrease the demand for the late forms of this rare headgear. The same cannot be said of the original First Empire examples, which tend to sell in the region of USD 1,500/2,000.

Notes:

1 The last vestige of armour to survive the advent of the 18th century was the breastplate used by the cavalry. Sappers occasionally made use of heavy siege helmets of burgonet form but the patterns of these helmets were by no means standardised. Cimarelli op.cit pg.9;

2 Wilkinson Frederick: Military Antiques pg. 34;

3 Balbi Da Correggio op.cit;

4 Including Frederick Wilkinson op.cit pg.38, and Wilkinson-Latham R..J; *Collecting Militaria* pg 43. The latter suggests that another reason hy the mitre hat was preferred to the grenadier cap was that the Tricorn hat could hindred the slinging of the musket over the shoulders.

5 Rankin Robert; *Military Headdress a pictorial history of military headgear from 1660 to 1914* pg 10;

6 Inalcik Halil: The Ottoman Empire, the Classical Age 1300-1600 plate 16;

7 In Germany and Russia;

8 Hollins David & Burn Jeffrey: *Austrian Grenadiers and Infantry 1788-1816.* Illustration on pg. 4 depicts a grenadier wearing his grenadier cap while four infantrymen wear raupenhelms. Illustration on pg. 30 illustrates four grenadiers wearing tall caps on a table with an infantryman wearing a Tricorn hat and three infantrymen wearing wide brimmed hats folded on one side. Vide also i18th century illustrations from Militar Pratica, 1740 published in Soares Branco Pedro: Coberturas Militares Portuguesas 1740-1918;

9 Kube Jan: Militaria, *A Study of German Helmets & Uniforms 1729-1918* p.33;

10 Tavard op.cit pg. 140;

11 Soares Branco Pedro op.cit pg.15;

12 Aprile's description of his headgear is corroborated by a painting of the period, Antoine Favray's Interior of saint John's Co-Cathedral;

13 Malta National Library Libr.Ms 466 fol. 59b;

14 Muller op.cit figures 165 and 166;

15 Muller op.cit figs. 230-280;

16 Wilkinson-Latham R..J op.cit pg. 44;

17 Kube op.cit p.33;

18 Such as the 1st and 2nd Foot Guard Regiment as well as the Tsar Alexander Guard Regiment No. 1;

19 Kube op.cit pp.34 & 42;

20 Kune op.cit p.38;

21 Muller op.cit figs. 267 & 268:

22 Wilkinsn-Latham R. J op.cit plates 68 & 69;

23 Tavard op.cit p.139;

24 Tavard op.cit pp.139 & 140;

25 Tavard op.cit p.145;

26 With an Imperial eagle in lieu of the grenade motif;

27 Without a helmet plate;

28 Tavard op.cit pp.145-149;

29 Tavard op.cit pp.150-151;

30 Tavard op.cit pp.151-152;

31 Wilkinson-Latham op.cit p.44;

32 Muller op.cit fig. 263;

33 Rankin op.cit p.87;

34 Lewis Waverly: US Military Headgear 1770-1880 fig 6;

35 Rankin op.cit fig. 147;

36 Wilkinson-Latham fig. 2;

37 Without fittings;

38 Carman W.Y: Head dresses of the British army/Cavalry p.15;

39 It was substituted with the Albert Pattern metal helmet only in 1842. Carman op.cit p.16;

40 Wilkinson-Latham op.cit fig.72;

41 Hollins David op.cit p.4;

42 Rankin Robert: The Official 1983 Price Guide to Military Collectibles p.178;

43 Hollins David op.cit p.24;

44 Rankin op.cit fig.149;

45 Wallis & Wallis Catalogue Sale 349 Lot 211;

46 Tavard Christian: *Casques et Coiffures Militaires Francais De' l'epoque gauloise a' la seconde guerre mondiale* p. 121

47 Rankin op.cit p.10;

48 And was known as the 'Monmouth cock'. Wilkinson-Latham p.42;

49 Rankin op.cit fig.5;

50 Tavard op.citp.123;

51 Tavard op.cit p.122;

52 Rankin op.cit p.10;

53 Lewis op. cit figs.7, 8 and 9;

54 Soares Branco: op.cit p.79;

55 Tavard op.cit p.37;

56 Carman op.cit fig. 17;

57 Carman op.cit p.46;

58 Tavard op.cit pp.38-41;

59 The crest of French helmets was fashioned in the form of a fierce medusa head with gorged eyes and serpent hair but the image of this gorgon was confused with that of the friendlier goddess Minerva.

60 Wilkinson-Latham op.cit p.56;

61 Knotel print; Chevauleger von Regiment Leningen;

62 Muller op.cit fig. 298;

[63] Kube op.cit p.36;
[64] Kube op.cit p.36;
[65] Tavard op.cit p. 209;
[66] Kube op.cit p.35;
[67] Tavard op.cit pp.162, 163;
[68] Rankin op.cit p.90;
[69] The overseas cap is no more than a twentieth century shrunken version of the *Zipfelmutze* worn in the eighteenth century;
[70] Hollins D op.cit: plate E;
[71] Macdonald John: Great Battlefields of the World p. 61;
[72] Departure of the Royal Guard to the Scottish Campaign 1745 (Artist Unknown) published in Orlandi Enzo: Gli Hannover Windsor p.25;
[73] Tavard op.cit p.147;
[74] Period painting published in Bull op.cit pp.136 and 137;
[75] Bull op.cit p.132;
[76] Hollins op.cit pp.61 and 62;
[77] Dorn Gunther & Engalmann Joachim: The Cavalry Regiments of Frederick the Great;
[78] Dorn Gunther & Engalmann Joachim: The Infantry Regiments of Frederick the Great;
[79] 1709 - published in Snellgrove LE: The early modern age pp. 237, 238; This tapestry also illustrates a British soldier wearing a mitre hat
[80] 1704;
[81] Wallis & Wallis Sale 396 Lot 694;
[82] Sotheby's March Sale: 'To Arms! Unifoms Painted Knapshaks, Canteens, Hat plates and Other Related Accoutrements;
[83] Including a considerable number of US Casquettes, Shakos and other artefacts from the American War of Independence and the War of 1812;
[84] Solis-Cohen Lita: The William Guthman Sale published in the Maine Antique Digest 2003;
[85] Lot 13879;
[86] Lot 11599;
[87] www.lehussard.fr reference 1214;
[88] Ebay Item#2176795490;
[89] Auction 45
[90] Lot 40
[91] Lot 37
[92] Lots 38 & 39
[93] Lot 24
[94] Kent Sales Sale 205 Lot 200;

5

The Neo-Classical Helmets of the Napoleonic Age

From Friedland to the Somme

"*Dragons que Rome eût pris pour des légionnaires,*
Cuirassiers, canonniers qui tránaient des tonnerres...."

Victor-Marie Hugo (1802–1885)
L'Expiation

The studies of behavioural psychologists suggest that headgear fitted with a long horsehair mane tends to grant its wearer an imposing and intimidating warlike appearance.[1] At a time when battles were fought with arms blanche and engagements were decided by hand-to-hand combat, the martial appearance of the soldier was quintessential, and the success of the crested Casques a La Romane of the eighteenth century is easily appreciated. Casques a La Romane remained in use up to the late eighteenth century, when in Napoleonic France the Casquette blossomed into the neo-classical helmet depicted by David and Detaille. This helmet influenced the development of the cavalry helmet throughout the nineteenth century and the first quarter of the twentieth century.

5.1 Casque a' Cimier and Casque a' Chenille

"*...le casque poli,*
Tous, ceux de Friedland et ceux de Rivoli,
Comprenant qu'ils allaient mourir dans cette fête,
Saluèrent leur dieu, debout dans la tempête."

Victor-Marie Hugo (1802–1885)
L'Expiation

The French light cavalry made extensive use of the crested Casquette from 1763 to 1786. In 1786 the popularity of this type of helmet was consolidated with a new model, the 1786 Model. The 1786 Model was a leather helmet that was based on the Casquette. It was fitted with a fur turban and a fur covered brim. It was also fitted with a low crest fashioned in the shape of a lion's head. This helmet was worn up to 1791, when it was replaced with the 1791 Model. The M1791 included some innovations on the M1786. The M1791 was a stronger helmet than the M1786. The M1791 had a metal skull and a brass crest fashioned in the shape of the characteristic French Medusa head. It retained the fur turban of the M1786, but it had a plain leather brim.

A very interesting French M 1804 pattern cuirassier helmet and M 1812 cuirass issued as a Belgian M 1831 cuirassier armour. (Private Collection - Malta) Photo by M Bonnici

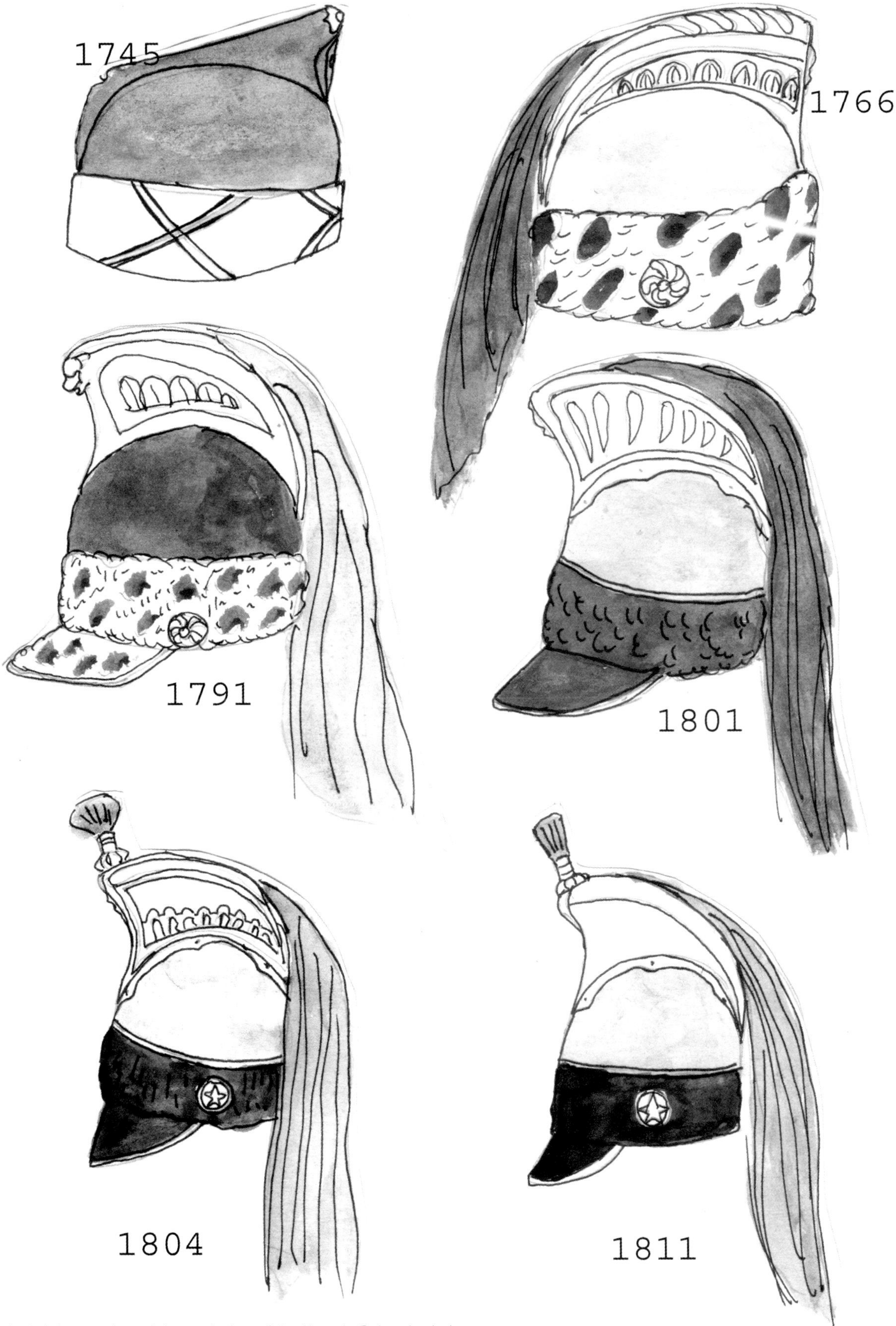

Artist's impression of the evolution of the French Cuirassier helmet.

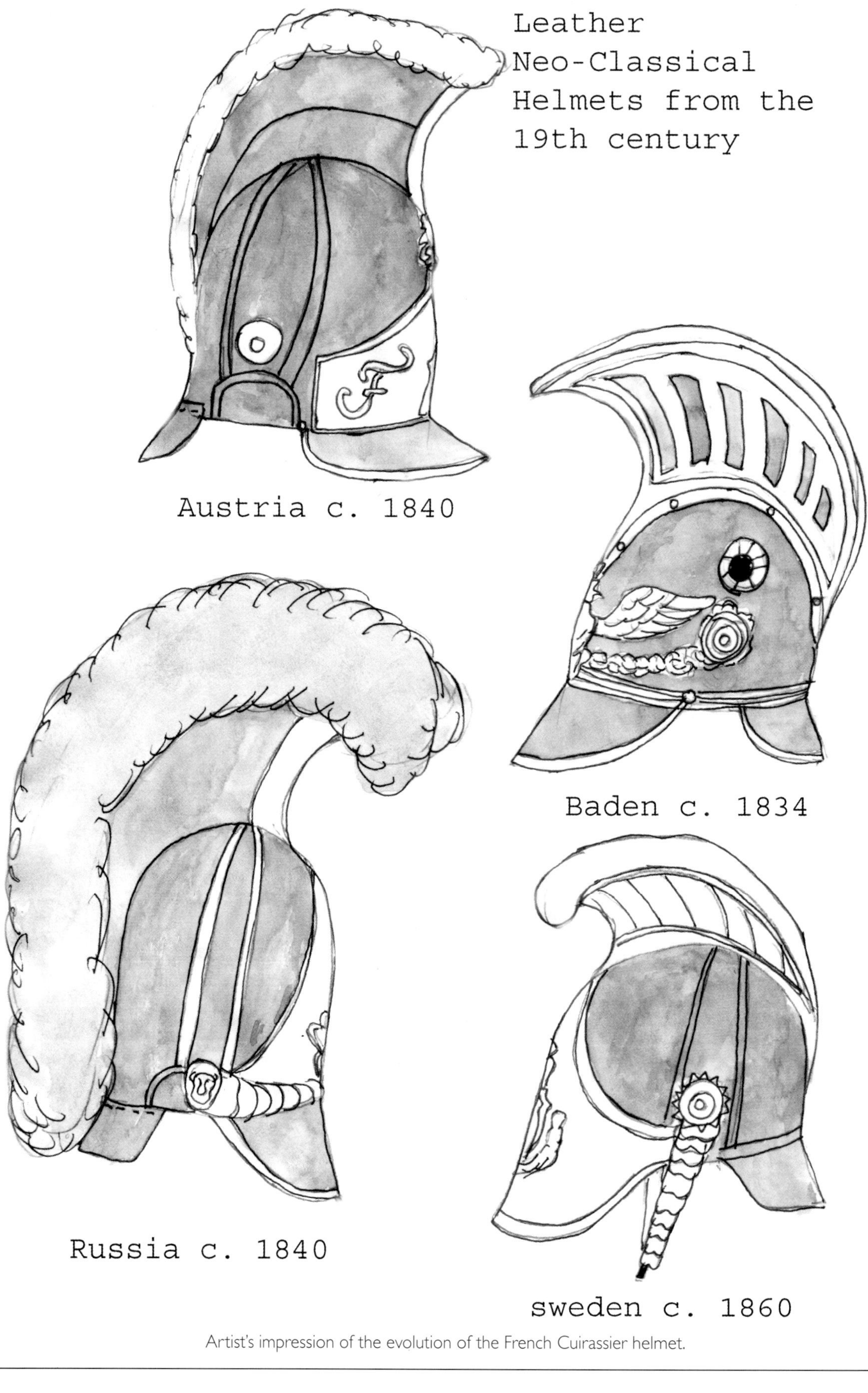

Artist's impression of the evolution of the French Cuirassier helmet.

This helmet evolved from the Napoleonic helmet after Napoleon raised the legendary 'cavalerie cuirassier.' In October 1801 Generals Kellermann and Bourcier equipped the first cuirassier regiment with a metal helmet that started a military tradition. It was made of a metal skull that was slightly higher than the M1791 and a taller brass crest. Another point of difference from the earlier helmet was the colour of the fur turban. In the M1801 the turban was made of black fur. In 1804 a slightly different version was issued. The M1804 had a higher and larger skull, and a fluffy bearskin turban.[2]

Further changes to the French Cuirassier helmet were made in 1811 when a cheaper 'economy' cuirassier helmet was issued. In that year Major Bardin introduced the Model 1811, a helmet with a plain crest that had no embossing.[3]

Helmets of a similar concept to the cuirassier helmet were worn by the Carabiniers. The latter were equipped with Casques a chenille in 1809. The Casque a chenille was a helmet of Greek inspiration. A metal band stamped with Napoleon's cypher replaced the turban of the cuirassier helmet. The helmet had a pointed metal brim and a flared rear brim. The brass comb was surmounted by a scarlet fur crest combed forwards. Similar helmets were worn by the Sapeurs du Genie de la Garde Imperiale. The latter's helmet was fitted with a large helmet plate with widespread wings instead of the metal band. In 1825 a similar helmet was issued to the cuirassiers. The return of the royalists to France in 1814 started a new helmet fashion. The cavalry of the Imperial Guard was substituted by the Chevau-legers de la Garde du Roi, and this elite unit was equipped with tall leather helmets fashioned in the pattern of the 1809 Casque a chenille. These helmets are the finest helmets of all time. These helmets were lavishly embellished with finely moulded brass fittings and delicately embossed combs.

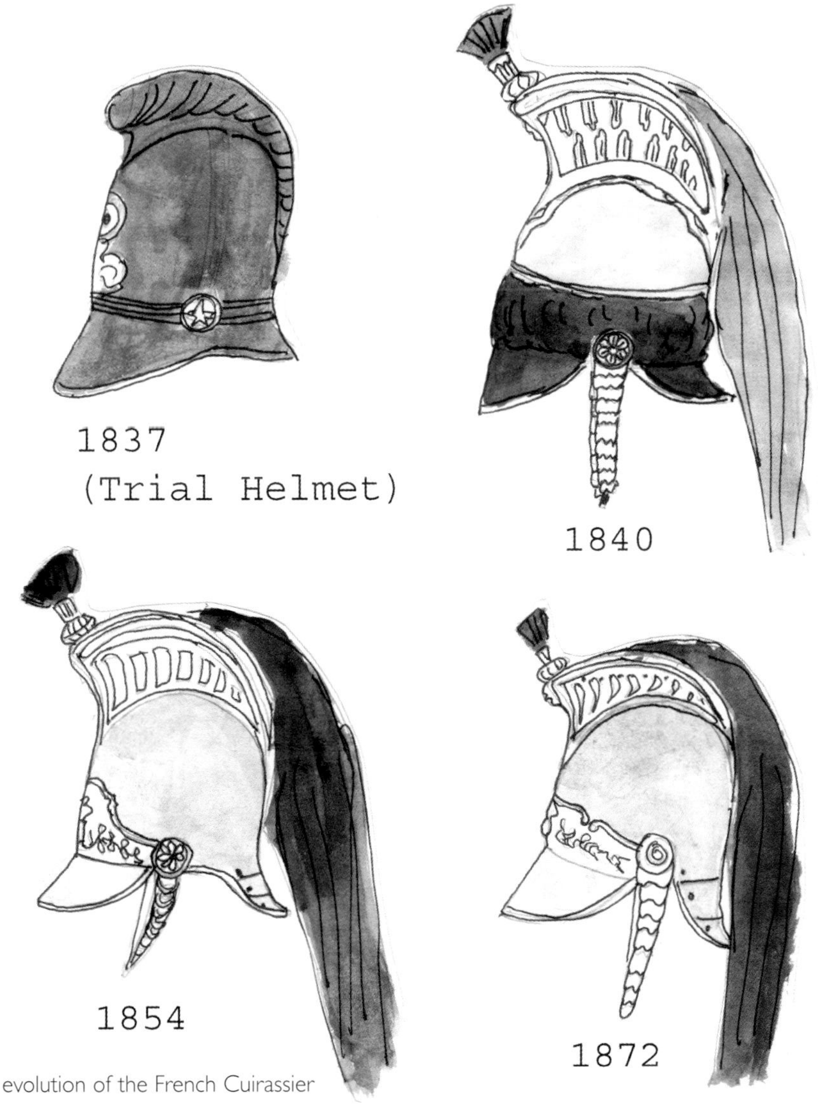

Artist's impression of the evolution of the French Cuirassier helmet.

A brass neo-classical helmet of the Garde Municipale De Paris a cheval c. 1830-1838, a very rare helmet. *(Courtesy of J P Morange)*

Side view of the Garde Municipale helmet.

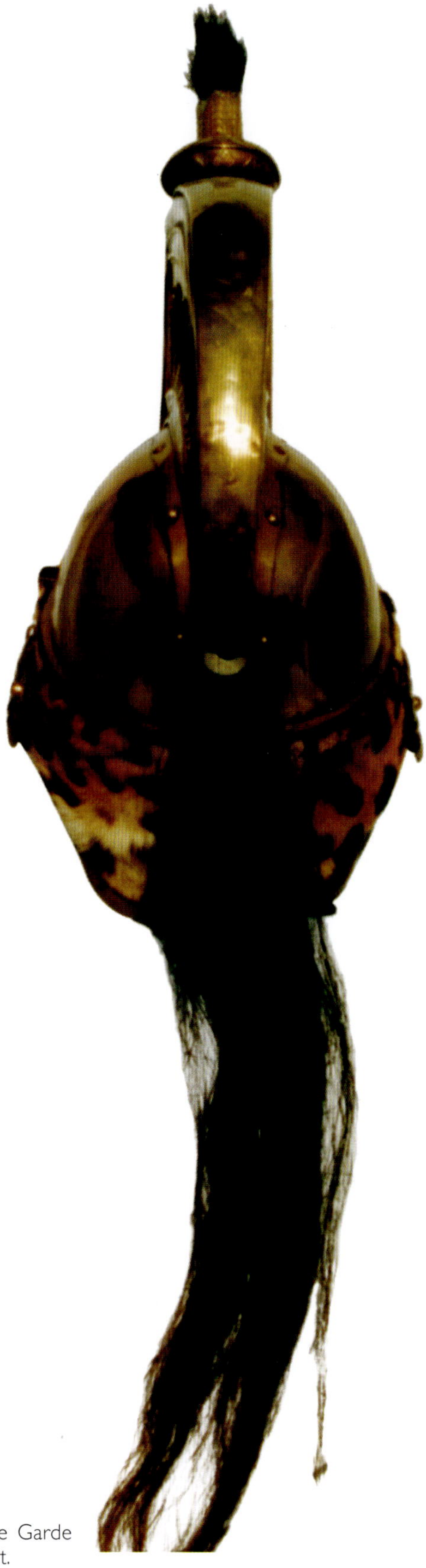

Rear view of the Garde Municipale helmet.

The French cuirassiers shed their M1825s in favour of a cuirassier helmet of Napoleonic inspiration in 1845. The M1845 was a facsimile of the M1804 with some minor changes. The M1845 had a metal brim and short flared skirt. In 1858 this helmet was replaced with a quasi-identical helmet with a lower dome.[4]

The popularity of this Napoleonic style helmet predictably depended on the rise and fall of the Napoleonic regime. The M1858 helmet fell in disuse with the restoration of the republic, when a new cuirassier helmet, the Model 1872—a reduced version of the republican M1825—came into being.

This helmet fell in disuse in 1915. The Gendarmerie Nationale adopted a brass and white metal version of the steel and brass M1872 in 1913.

The latter helmet fell in disuse a few years later.[5]

A French Model 1858 Cuirassier helmet converted to a Fireman's helmet. Published in Treasures of Malta Vol VIII No 3. (Private collection Malta) Photo by M Bonnici

Below left: This cuirassier helmet was converted into a fine helmet with the clipping and shortening of the comb. Below right: Rear view of the helmet. Photos by M Bonnici

A fine French M 1872 pattern helmet to the dragoons/cuirassiers, the last of the cuirassier style helmets issued after the fall of Napoleon III. It appears to be a relic from World War One. (Private collection - Malta) Photo by M Bonnici

Close up of the grenade style helmet plate, a characteristic that was present in a few napoleonic helmets, but was mainly used on republican helmets. Photo by M Bonnici

Close up of the Medusa face finial on the helmet's comb. Photo by M Bonnici

A brass and white metal French gendarmarie a cheval helmet c. 1913-1915. The last of the neo-classical pattern helmets. Published in Militaria International Magazine October 1999. (Private collection - Malta) Photo by M Bonnici

Side view of the M 1913 Gendarmarie helmet plate. It has a lower profile to that of certain other French Neo-classical helmets, and bears some resemblance to the Adrian helmet. Photo by M Bonnici

Close up of the Medusa face finial on the helmet's comb. Photo by M Bonnici

Rear view of the Gendarmarie helmet. Note the armourer's mark of Franck et ses fils. Photo by M Bonnici

Close up of the Gendarmarie helmet's chinscale bosses. They were inspired from 1814 *Maison Du Roi's* helmets. Photo by M Bonnici

Close up of the Gendarmarie helmet's plate. Photo by M Bonnici

5.2 Casques a' Cimier on the Continent and Beyond

"Oh, there is a tale about Buzulukov—simply lovely!' cried Petritsky. 'You know his weakness for balls, and he never misses a single court ball. He went to a big ball in a new helmet. Have you seen the new helmets? Very nice, lighter. Well, so he's standingÖ. No, I say, do listen."

Leo Tolstoy
Anna Karenin
1875-1877

The combed helmet surmounted by the imposing horsehair mane became one of the most evocative artefacts reminiscing the charge of Napoleon's splendid cuirassiers at the battle of Waterloo. The helmet pattern was so successful that Napoleon's friends and foes alike started to use the 'Napoleonic' helmet. One of the first nations that adopted the helmet was France's neighbour, the newly constituted Belgian State.[6] Belgium's first cuirassiers were equipped with French M1804 type[7] helmets produced by the same armourers that produced the helmets worn by Napoleon's legendary cuirassiers.[8] These helmets were denominated M1830. It appears that Belgian cuirassiers also wore leather versions of this helmet.[9] The Belgium M1830 was rendered obsolete in1843[10] when it was replaced by an all-metal crested helmet based on the M1830.

A Model 1808 helmet Russian Cuirassiers' helmet with horsehair bush.

This page and following: A fine Napoleonic helmet and cuirass of the Russian line Cuirassiers. *(Courtesy of Hermann Historica)*

Belgian M 1831 Cuirassier helmet with heavy iron skull, leather turban and peaks, and a tall neo-classical style brass comb. (Private collection Malta) Photo by M Bonnici

Side view of the same helmet. Fittings are identical to those found on most First Empire Helmets. Photo by M Bonnici

Close up of the Versailles armourer's mark on the rear of the helmet's bowl. This fine helmet is dated 1831. Photo by M Bonnici

Close up of the armourer's mark on the comb. It is the mark of Fonsons of Brussels, an armourer that produced armour for Napoleon I. Photo by M Bonnici

The 1843 helmet did away with the classic turban in favour of an impressive large helmet embossed in the shape of a lion's face.

Even the German states employed French cavalry style helmets. Although Prussia used distinctly Germanic tall leather neo-classical helmets, such as the M1808 line cuirassier helmet,[11] other German states used facsimiles of French cavalry crested helmets. In between 1810 and 1826 the Garde du Corps of Saschen-Gotha-Altenburg[12] wore a French M1804 helmet with back peak. In 1814 Bavaria's Garde Du Corps wore an almost identical helmet to the1809 Casque a chenille.

This leather and metal helmet was replaced with a slightly lower metal helmet in 1833. The Bavarian M1833[13] was worn up to 1842[14] when the size of the crest was lowered. This helmet fell in disuse in 1873 when it was replaced with a light sheet iron helmet of simpler form.[15]

Other German states wore helmets that combined elements of the late leather casquette and the Napoleonic metal helmet. The infantry regiments of Sachsen-Coburg-Gotha wore leather helmets that had a substantial brass helmet plate and a woven fur/hair comb in 1849.[16] A helmet of a similar construction with a brass-reinforced comb was worn in Saxony from 1849 to 1867,[17] and in Denmark throughout the nineteenth century. These helmets were known as 'raupenhelms,' and were particularly popular in Bavaria with the infantry regiments. Raupenhelms appeared in Bavaria in 1815 with the Model 1815, a large leather helmet surmounted with a fur crest.[18] The size of this helmet was reduced with the M1845,[19] and further reduced with the models 1868[20] and 1873.[21] The M1873 helmets were replaced by the Picklehaube, a helmet that would dominate German uniforms and traditions for years to come.

Even the Austro-Hungarian Empire and the emerging Italian state were swept up in the craze for the Napoleonic helmet. Austrian dragoons started to wear leather and metal neo-classical helmets in the first quarter of the 19th century, and continued to wear neo-classical helmets up to the First World War,[22] when a special greyed matted version of the neo-classical helmet was issued to cavalrymen on active service. A very similar, slightly more ornate helmet fitted with the Savoy cross was worn in Italy between 1864 and 1930.[23] Prior to

Prussia c. 1809

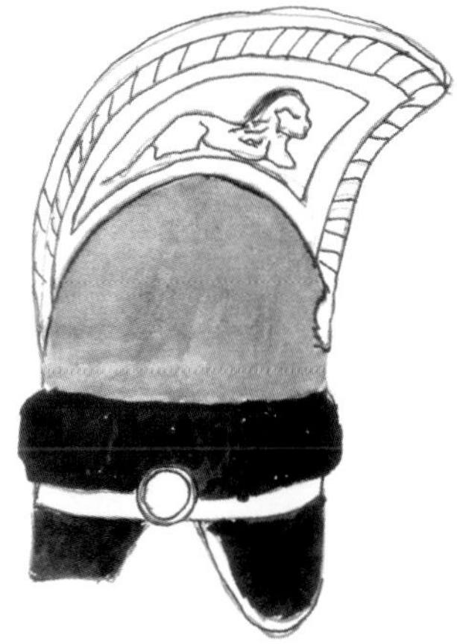

Saschen-Gotha-Altenburg
1810-26

Austria c. 1860

An Artist's impression of a series of neo-classical helmets worn in the German states.

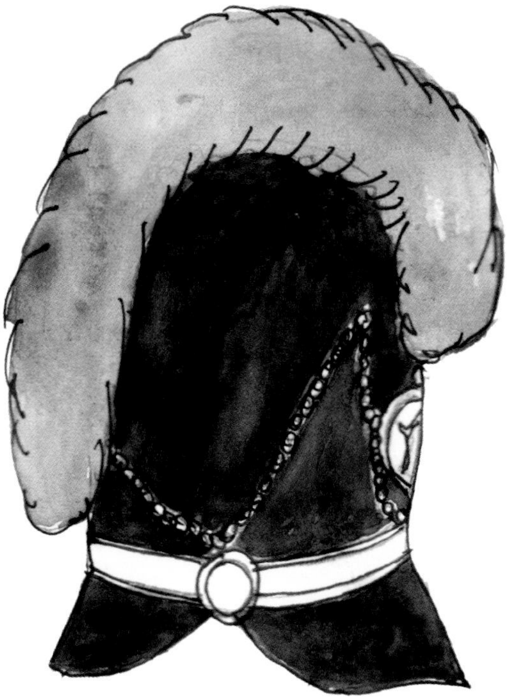

1815

1845

1848

1868

An Artist's impression of the evolution of the Bavarian raupenhelm.

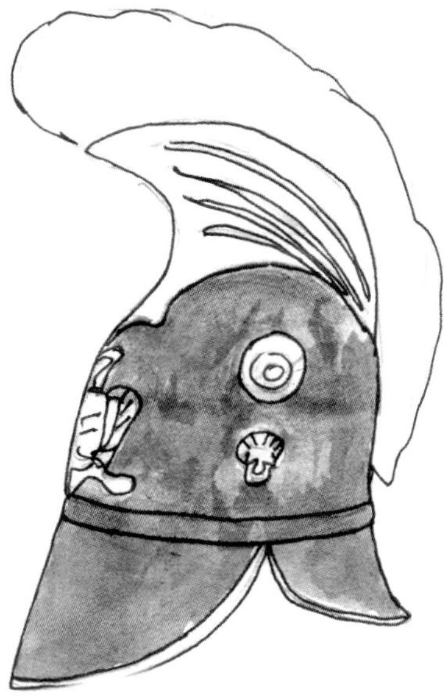

Bavaria
Home Guard
M1848

Switzerland
c.1850-1890

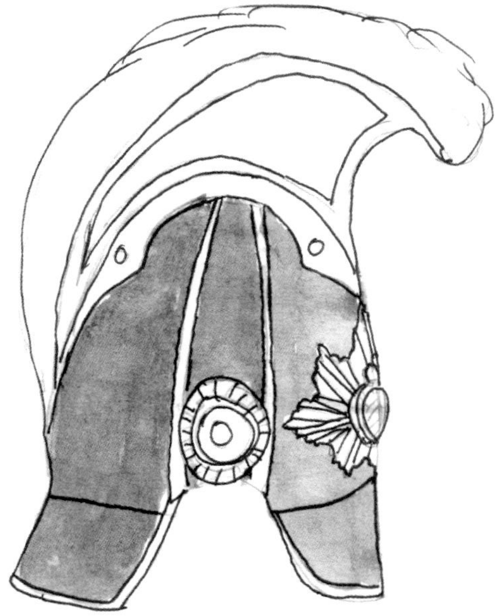

Saxony
c.1867-1890

An Artist's impression of the evolution of the Bavarian raupenhelm.

unification, the dragoons of the National Guard used a helmet with a French style turban and a British type plume between 1848 and 1864.[24]

An even more ostentatious helmet was worn in Spain in the first years of the 20th century.[25]

Portugal experienced a brief interlude with the use of a neo-classical style helmet in the last years of the monarchy. In 1874 Portugal adopted a very similar helmet to the French M1872. The medusa head was replaced by a lion's head, and the royal cipher took the place of the flaming grenade, but the helmets of the Guarda Municipal De Lisboa were quasi-identical to those worn by the French cuirassiers and dragoons of the period.[26]

The popularity of the neo-classical helmet spread outside of the European continent. The Persian Royal Guard used helmets similar to those worn by the French Cent Garde[27] in the mid-19th century, and certain American militia units, such as the Philadelphia City Troop, used helmets that were based on the helmets of the Maison Du Roi.[28]

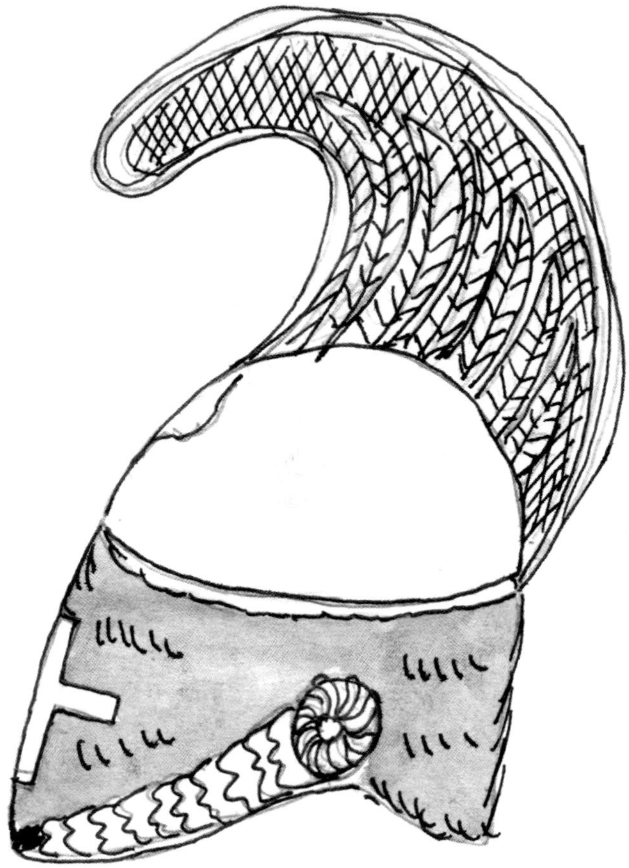

Italian M1864

An Artist's impression of an Italian Model 1864 cavalry helmet.

An officer's helmet of the Milanese Guardia Nazionale c. 1848. *(Courtesy of Hermann Historica)*

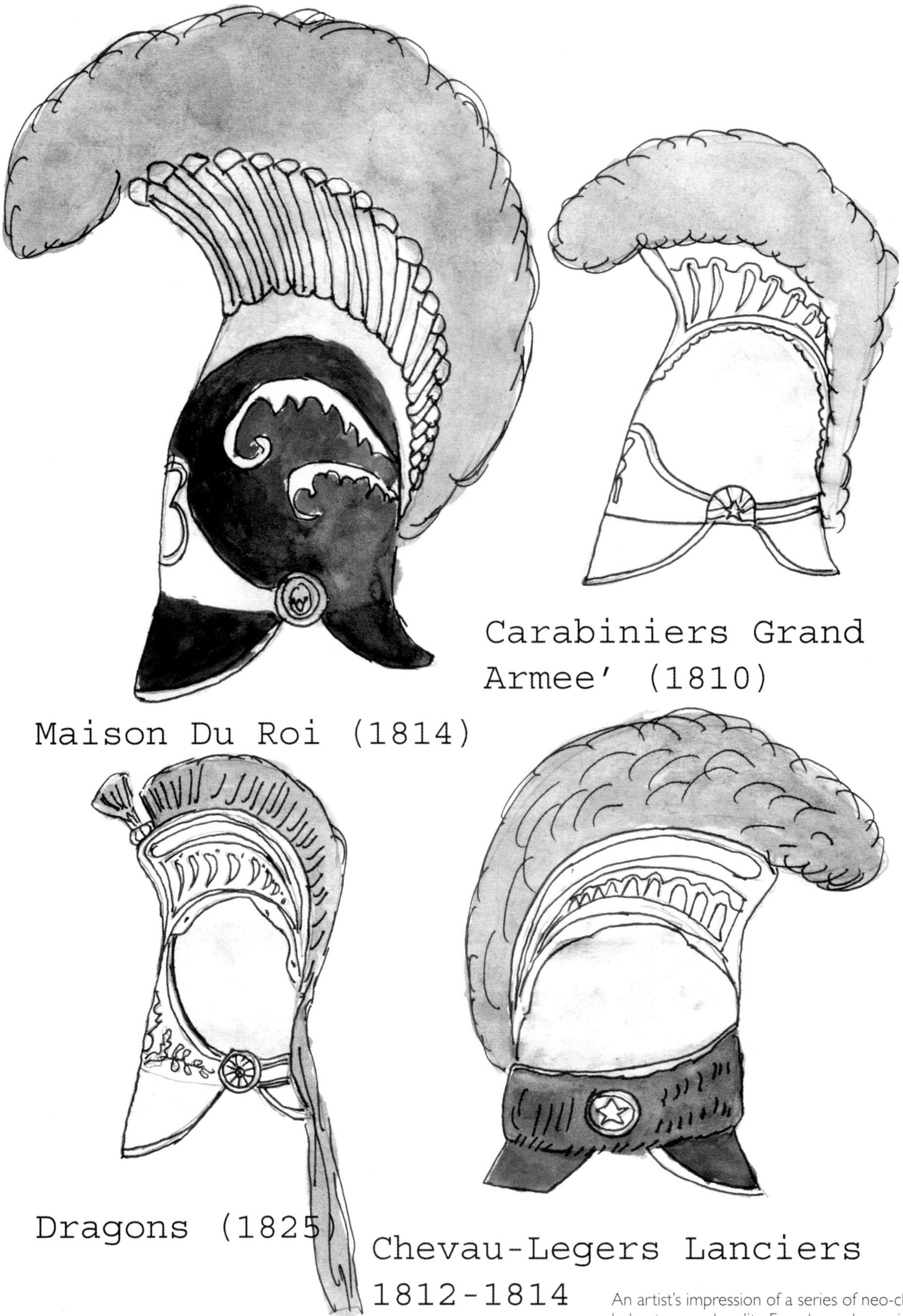

An artist's impression of a series of neo-classical helmets worn by elite French cavalry regiments.

5.3 The Neo-Classical helmet in Great Britain

"*I noticed the valorous cuirassiers many times. They used to arrive right under the squares and withdraw. At times they used to gallop in our midst but they did not attack, it would have been a certain destruction....*"

Lieutenant Jackson
1815[29]

Although Great Britain was Napoleon's greatest enemy, the British army was one of the armies that plagiarised Le Grande Armee's uniforms and traditions. Napoleon's troops had left a great impression on the British army at Waterloo. Five years after the bloodbath at Waterloo, Britain's heavy cavalry started to wear a cuirass similar to that worn by their foe. The Napoleonic influence was by no means limited to the cuirass, but extended itself to headgear.

One of the first regiments to wear an offshoot of the French M1804 was the regiment of the Life Guards. In 1812[30] they were equipped with a leather helmet of M1804 shape, with bombe' a' la minerve fitted with a large helmet plate, along with the Georgian cipher and a brass turban. Two rare specimens of an officer's and trooper's versions of this helmet are conserved at the Household Cavalry Museum.[31] The trooper's version possessed some distinctly 'British' characteristics that were retained on British helmets for the next century or so. These were the chin scale bosses fashioned in the shape of lions' heads[32] and the 'v-shaped' down pointing brim.[33] If the French M1791 and M1804 were a source of inspiration for the M1812, the flamboyant helmets of the Garde de Maison Du Roi were the source of inspiration for the M1817, a helmet chosen by the Prince Regent. The M1817 helmet was a helmet of bright polished steel with appliqué gilt brass ornaments, with

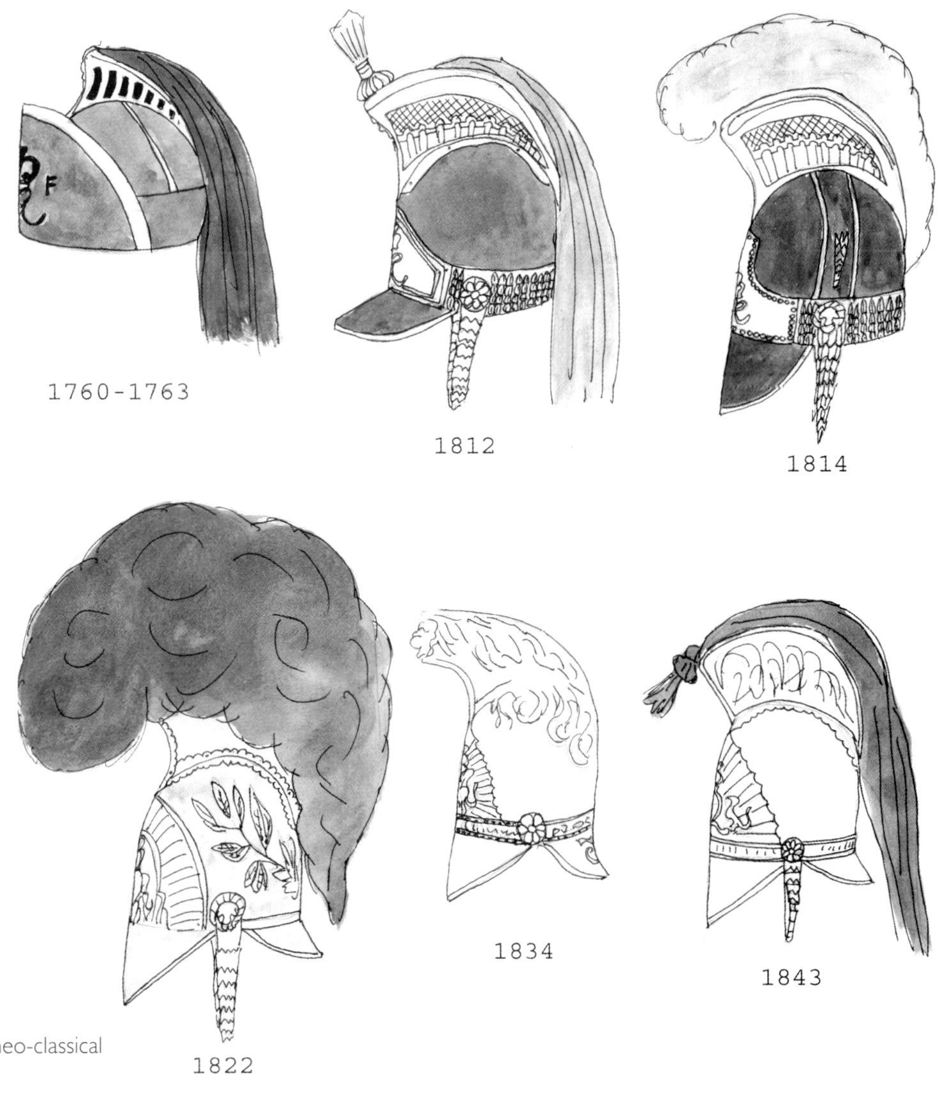

An Artist's impression of early British neo-classical helmets.

a tall brass comb adorned by a large bearskin crest. In dress regulations this helmet was referred to as the 'Roman' helmet. A massive bearskin cap 'fourteen inches deep' substituted this helmet in 1833. A year later dragoon guards and dragoons received another 'Roman' helmet denominated the 1834 pattern. This was an all brass helmet, having a skull that had a similar shape to that of the M1817. The differences from the M1817 consisted in the appliqué decorations and the crest. Unlike the M1817, the M1833[34] was not embellished with appliqué laurel leaves, but with curling foliage. The crest of the M1834 could be worn in conjunction with a bearskin crest identical to that used on the M1817, but the 1834 was also worn with a gilt brass comb in the shape of a rampant lion's head.[35]

Prince Albert approved another similar helmet in 1842. Although it was in use by the Royal Horse Guards as early as 1842, it was officially authorised for the dragoon guards only in 1847.[36] The leading authorities on British heavy cavalry helmets[37] and British uniforms and traditions[38] suggest that the British copied this helmet from the 'Piklehaube type' helmets worn by the Russians and Prussians at the time. The author begs to differ with their point of view. Certainly the British did not copy the M1842 from the Russians, because the Russians started using a 'Picklehaube type' helmet only on May 9, 1844,[39] two years after the British M1842 made its appearance. Similarly, the author does not believe that the British M1842 was copied from the Prussian Picklehaube for a number of

A rare 1847 style helmet of the Afghanistan Horse Artillery. (Private collection - Malta) Photo by M Bonnici

reasons. The first Picklehaube appeared in Prussia on October 23, 1842,[40] and although the exact date when the first British M1842 went into production is unknown, it is unlikely that ideas could travel so fast in those days. More importantly, Germany's first Picklehaube looked very different from the British M1842. Unlike the German leather spiked helmet, the British M1842 was not made of leather.[41] Similarly, the shape was different. The first Prussian Picklehaube had a characteristic square peak, a characteristic spike,[42] and a cruciform spike base. None of these elements are found in the British M1842. The British M1842 consisted of a white metal[43]/silver[44] or brass[45]/gilt copper[46] skull constructed in two pieces and embellished with a foliate helmet plate and appliqué strands of acanthus foliage,[47] with a front peak—which traced its shape back to the British heavy cavalry M1817—and a back peak that traced its shape to the British Dragoon Guards M1817. The only common denominator between the Prussian M1842 and British M1842 was the type of plume holder; an important break with the French neo-classical combed helmet tradition.

Side view of the same helmet. The helmet appears to have been vandalised. The horsehair plume is missing, and the plume holder is broken. Photo by M Bonnici

Above: Close up of an M 1847 style helmet of the Afghanistan Horse Artillery. Photo by M Bonnici

Left: Close up of the helmet plate of an M 1847 helmet of the Afghanistan Horse Artillery (copper spray missing). It is of a markedly inferior quality compared to the helmet plates of M 1847 British cavalry helmets. Photo by M Bonnici

The M1842 was substituted by a simpler 'economy issue' helmet in April 1871. This helmet was of a similar construction to the M1842, but its ornamentation was more austere. The strands of acanthus foliage were reduced to the bare minimum; only a strand around the crown was left. A simpler socket shaped plume holder was adopted, and a range of new star-shaped helmet plates was used. The plume holder base of the M1871 helmet had a cruciform shape. The regiments of the dragoon guards and some of the yeomanry regiments wore this helmet.[48] The Horse Guards wore a slightly more elaborate helmet. The M1871 fell officially in disuse in 1939 when full dress was abolished,[49] but it is still used by bandsmen of the 4th/7th Royal Dragoon Guards.[50] The Horse Guards continue to wear this helmet today.

Left: Artist's impression of an 1847 pattern British cavalry helmet. Right: Artist's impression of an 1871 pattern British cavalry helmet.

Variants of the M1842 and M1871 were worn by some British Yeomanry regiments. The Norfolk (King's Own) Yeomanry Regiment wore a leather version (with round brim) of the M1871.[51] The Staffordshire Yeomanry wore an M1842 style helmet of black japanned metal.[52] The Duke of Lancaster's Own Yeomanry wore a similar helmet in japanned metal, but the brim of their helmet lacked the laurel leaf strands. Other voluntary regiments, such as the Duke of Manchester's Mounted Rifle Volunteers, wore leather helmets similar to the M1812, but of a cheaper construction.[53] The Fife Mounted Rifles wore a leather helmet that combined elements of the latter helmet and were fitted with a queer plume holder of M1871 inspiration.[54]

Rear view of an M 1847 British helmet and an M 1871 British helmet. The M 1874 is more heavily decorated than the M 1871. (Private Collection Malta) Photo by M Bonnici

5.4 Battles Fought in Neo-Classical helmets

"...It was a moving scene, I had tears in my eyes. A group of cuirassiers galloped beside me. Their cuirasses shone. I saw them charge and never saw them again...."

A French Colonel
1815[55]

Surviving specimens of the French M1801, M1804, and M1811 cuirassier helmets in museums prove that these helmets were used extensively on the Napoleonic battlefield.[56] Decisive cavalry engagements, such as those at Waterloo and Friedland, were fought with these neo-classical helmets. On the other hand, the M1845 French cuirassier helmet was worn in the Crimean War,[57] and the M1858 was worn in the Mexican War and the Franco-Prussian War. M1874 helmets—the last helmets of a Napoleonic pattern—were worn in the first part of the First World War, when the shiny metal components of the M1874 helmets were covered by cloth covers.[58]

The British Heavy Cavalry and the Dragoons fought the Crimean War in the M1842 pattern helmets.[59] A British officer who fought in the Crimean War spoke very well of his M1842 pattern helmet. He described his helmet as being comfortable. It appears that the M1842 was also worn during the Indian Mutiny, when it was fitted with a puggree wrap. It appears that this helmet was worn at the battles of Balaklava, Sebastopol, and Lucknow.[60] The M1871 appears to have been, like the M1817 and M1834 helmets, only a parade helmet, because all the battles fought by Queen Victoria's troops during the M1871's term of service were fought in Pith Helmets.[61]

A Bavarian Model 1842 Cuirassier helmet (c. 1848-1851). *(Courtesy of Johan Somers)*

5.5 Collecting

The helmets described in this chapter represent the most interesting area of collecting. The neo-classical helmets of the 19th century combine aesthetic beauty with historic importance. Most of the helmets described in this chapter regularly appear on the collector's market, but they fetch prices that are commensurate with their visual appeal and history.

In 2003 a Model 1812 French Cuirassier helmet was sold in an Internet auction at USD 5,655. This helmet received no less than twenty bids.[62] First empire helmets rarely appear for sale, and there is always a lot of interest in these early examples. Unfortunately, the appeal of these helmets inspired the creation of numerous fakes. Modern forgeries lack the detailed artwork of the originals, and the light 'white' metals used in these replicas make these forgeries identifiable. The same cannot be said of the tailored Belgian M1830 helmets. Belgian M1830s are highly collectable in their own right, but it appears that some unscrupulous sellers are removing the back peaks of these helmets and offering them for sale as First Empire helmets. One can tell the difference between an M1804 First Empire helmet and an M1830 from the shape of the visor, the sign of tears near the back of the turban (the M1830 has a rounded visor) and, more visibly, the inscriptions on the rear crown (all Belgian M1830s seen by the author had date marks ranging from 1830 to 1840). One cannot exclude that future fakers may seek to erase the date mark, so a helmet with an erased mark should always be looked upon with suspicion. The purchase of a First Empire helmet always implies a significant financial investment, so First Empire helmets should always be purchased from experienced and specialised outlets.

Even Second Empire helmets and republican neo-classical helmets are being sold for significant prices. In 2003 a French dealer was offering for sale a fine brass helmet Model 1858 of an officer of the dragoons at around USD 4,000.[63] A British dealer offered a trooper's version of the same helmet at USD 2,392.[64] In the same year another British dealer offered a complete French M1913 Gendarmerie helmet at USD 1,650,[65] while a colleague of his offered a French cuirassier trooper's helmet at USD 952.[66] Even the cloth covers for M1874s are selling at USD 297.[67]

Early British cavalry head dress and the legendary and rare helmets of the Garde de Maison Du Roi are exceptionally expensive. In 2003 a British Officers' M1818-1834 was offered for sale in an internet auction with buy it now facility at USD 16,000.[68] In 2003 the same seller offered a British Officer's 1834 Helmet of the Royal Dragoons and a French Officer's helmet of the Mosqetaries Du Roi with buy it now facility at USD 15,500[69] and USD 23,500,[70] respectively. In 2003 an 1834 pattern Royal Dragoon Guards helmet was being offered for sale at USD 11,520,[71] while an 1847 Cavalry Helmet of the Royal Dragoon Guards was being offered for sale at USD 5,600.[72] The same dealer offered a Household Cavalry 1816 helmet at USD 13,600[73] and an 1834 pattern cavalry officer helmet of the Royal Dragoon Guards for the same price.[74] Even

These three helmets illustrate three different stages in the evolution of the French helmet, 1804-1913. (Private Collection Malta.) Photo by M Bonnici

1871 pattern Yeomanry helmets tend to be expensive. A British dealer offered a British Montgomeryshire 1871 Pattern trooper's helmet at USD 1,938.[75] Another British dealer offered an 1871 Pattern helmet of the Essex Yeomanry at USD 1,589.[76] The author heard of fake British 1816 and 1834 helmets, but he did not have the opportunity to investigate the matter further. The author noted some replica 1871 patterns being offered for sale, but these replicas had incorrect patterns and poor quality embellishments. Composite helmets (helmets assembled from original spare parts) are a much more concerning issue, as the components used in these helmets are original in every respect. At times it is impossible to identify a composite helmet, but extra holes on a helmet crown or enlarged mounting holes are indicative of tampering of some form or another. The author also heard of old M1842 and M1871 helmets being re-lined with modern liners used on household cavalry helmets. These should be easily recognised, as old helmets had leather linings for the ranks and silk lining for the officers, while modern helmets have light coloured chamois linings.

Even the prices of neo-classical helmets used by nations other than Britain and France tend to be costly. In 2003 an Australian dealer offered an 1843 Belgian helmet with the distinctive lion plate at USD 2,640.[77] In the same year an M1864-1930 Italian Dragoon helmet was being offered for sale atUSD 1,443,[78] while a famous German auctioneer valued a rare 1848 pattern at around USD 2,500.[79] In 2003 a British dealer offered an unusual Napoleonic style helmet of the Persian cavalry from the 1860s at USD 889.[80] Austrian dragoon helmets were always in great demand, and although they are not exceptionally rare, an officers' example was offered at USD 2,590 in 2003.[81]

A rare officer's helmet of the Saxony Reiterregimenter c. 1849-1867. *(Courtesy of Tom Berndt, Militaria International Magazine)*

The skull of a rare 1814 pattern Bavarian Garde Du Korps helmet refurbished c. 1914 as a re-enactment helmet and used as a theatre helmet in the 1940s. Private Collection Malta. Photo by M Bonnici

Inside view of the Napoleonic Bavarian helmet. Photo by M Bonnici

6

Shakos, Shapkas, and Busbies

From Waterloo to Balaclava

"The Infantry of the National Guard wear a blue frock with white buttons and epaulettes. The cap is dark blue, with a red lace band at the top. It is adorned in front with an eagle in gilded metal, and the number of the battalion. The uniform of the cavalry of the National Guard is blue, with a scarlet stripe on the pantaloons, and white epaulettes. The cap is is of the Polish style, with an eagle in front. These uniforms are, on the whole, quite plain, but serviceable and warlike."

Gleason's Pictorial Drawing Room Companion
1848

Most of the military headgear worn in the eighteenth century offered inadequate protection, and the need was felt for heavier headgear. Military tailors sought inspiration in Eastern Europe and created a flamboyant range of colourful military hats. These extravagant hats looked impressively splendid, but their extravagant appearance was not in keeping with their ballistic aspirations. It is truly amazing how these hats succeeded in surviving the opening years of the twentieth century.

A clipping from an 1848 newspaper that announces the latest French military fashion. (Private Collection Malta.) Photo by M Bonnici

A Napoleonic trooper's Shako of the Garde National regiment c. 1810-1815. *(Courtesy of J P Morange)*

Front view of the Garde D'Honneur Shako.

Above: Inside view of the Shako's liner. Below: Close up of the tailor's label of Monsieur Chardon, Fabricant de Shakos, rue de la Monnaie No 19, Paris.

Rear view of the Garde D'Honneur Shako.

In 1815 a special straight, tall cylindrical Shako, which did not have the bucket shape of the other models, was issued to the light cavalry. This Shako became known as the '*Rouleau pattern*'. It had various different features from the M1806, M1810, and M1812 Shakos besides the cylindrical shape. It was fitted with a loop and a button in lieu of a helmet plate, it had a straight rectangular peak, and had no chin-scales. This Shako remained unchanged till 1843.[8]

The M1812 Shako became obsolete with the fall of Napoleon. In September 1815 Louis XVIII decreed a new Shako pattern and, more importantly, a new range of helmet plates emblazoned with the fleur de lys. Although this Shako was intended to break with the Napoleonic tradition, it retained many features that were found in its earlier counterparts. The M1815 pattern was taller than earlier French Shakos. It was 22 centimetres tall, but the manner of construction (pressed felt) and its '*pot de fleur*' shape was similar to that of the M1812. The M1815 had a rectangular flat peak similar to that of the '*Rouleau pattern*' (earlier infantry Shakos had a down turned brim).

A very rare voltiguer 1815 pattern *'pot de Fleur'* Shako as worn after the fall of Napoleon. (Private collection - Malta) Photo by M Bonnici

The shape of the body of this Shako from the era of Louis Phillipe is quasi-identical to the shape of late First Empire Shakos. Photo by M Bonnici PCM

Close up of the tape of gold braid at the top of the Shako. The signs of oxidisation and wear are unequivocal signs of authenticity. Photo by M Bonnici

A close up of the helmet plate and the cockade. The plate is a first pattern plate *'au coque'* issued in 1830. Photo by M Bonnici

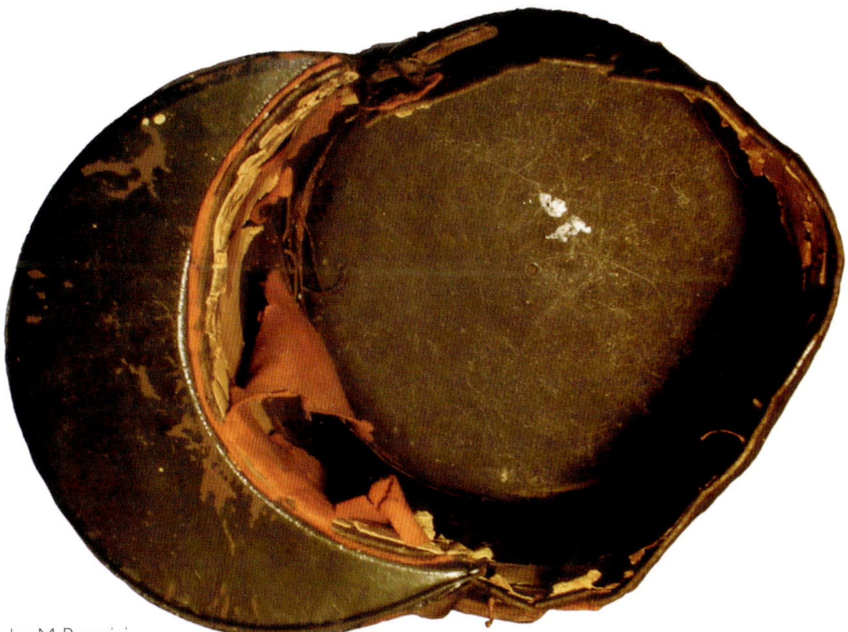

Inside view of the M 1815 Shako. Photo by M Bonnici

The British Army adopted a very similar Shako almost a month later.[9] The 1815 was a very light and fragile helmet built from pressed felt and cardboard, and in March 1825 it was substituted with another very similar helmet. The heavy Model 1825 was almost identical to the M1815 in shape, but the body of the M1825 was made of leather in lieu of cardboard. This likely explains why the M1825 is much more common than the M1815.

The July 1830 revolution and the crowning of Louise Phillipe marked another new range of helmet *'au coque'* plates adorned with a rooster. Mounting holes on M1815 and 1825 Shakos suggest that in 1830 the latter Shakos were stripped of their fleur de lys plates and fitted with cockerel plates. In July 1837 a new Shako model, the M1837, substituted the M1815 and the M1825. This Shako was significantly shorter than the earlier models at only 18.5 centimetres tall. Moreover, the 1837 was not a *'pot de fleur'* Shako, but a squattish *'tronconique'* (trunk shaped) Shako with straight sides. It is believed that this ugly Shako inspired British tailors to create the 1844 pattern.[10]

A scarce French officer's M 1840-46 pattern Shako of the 4th mobile infantry. (Private Collection Malta.) Photo by M Bonnici

Close up of the Shako's Louis Philippe period button. Photo by M Bonnici

Inner view of the French Shako's liner. Photo by M Bonnici

All the Shakos described above were considered too heavy for tropical service, and in 1840 the French army introduced a lightweight Shako style hat known as '*Casquette d'Afrique*,' a comfortable hat designed for the colonies. It was entirely made of felt and was fashioned in a totally innovative shape. Unlike the M1815 and the M1825 Shakos that widened out towards the top, the *Casquette d'Afrique* narrowed towards the top, decreasing the size and weight of the helmet. Whereas the M1815 and M1825 had a bucket shape, the *Casquette d'Afrique* looked like an upturned bucket. The innovative shape of the *Casquette d'Afrique* heralded a new Shako design. In September 1840 a sturdier version of the *Casquette d'Afrique* (the M1840-43) was issued to the troops. The M1840-43 had a narrower leather top than the M1815 and M1825, reducing the weight of the hat considerably. The M1840-43 Shako was also the first French Shako to look tilted forward, a style that was subsequently copied by British tailors.

Some versions of this Shako had another important feature; they were fitted with a pair of small round air vents.[11]

A French M 1840-43 Shako of the National Guard. (Private collection - Malta) Photo by M Bonnici

The French M 1840-43 Shako's profile was inspired from the 'tilted forward' profile of the *casquette d'Afrique*. Photo by M Bonnici

The establishment of the Second Empire in 1852[12] introduced changes in the uniforms and traditions of the French army.

Cockerel helmet and Shako plates '*plaque au coque*' were replaced with First Empire type eagle helmet plates '*plaque a l'aigle*.' The eagle helmet plates introduced in 1852 consisted of uncrowned eagles with closed wings, with their heads facing left. M1840-43 Shakos fitted with these plates were designated M1852.

A new range of eagle helmet plates was introduced in 1855. Unlike the earlier helmet plates, the plates introduced in 1855 featured crowned eagles. Shakos fitted with these plates were denominated M1855. A new Shako pattern, the M1856, was introduced a year later. The M1856 had the same tilted forward shape of its predecessors, but was slightly shorter (15 centimetres) than the M1840-43, and had a much more squattish appearance than its earlier counterpart. A heavier Shako of M1840-43 shape made entirely of leather was introduced four years later. This short Shako incorporated a pair of lateral chevron style strips on the sides reminiscent of the forgotten M1806. The 1860 Model was a shorter Shako than the M1856 at only around 12 centimetres tall.[13] Despite its several shortcomings, this headdress was rather popular, and it appears that France exported its M1860s to some militia units of the

Detail of the chinscale boss and chinscales of an M 1840-43 Shako of the volligeurs. Photo by M Bonnici

Inside view of the liner of the M1840-43 Shako. Photo by M Bonnici

Detail of the late *'au coque'* pattern helmet plate and the 1818 leather cockade. Photo by M Bonnici

Close up of the scarlet tape on the top of the M 1840-43 Shako. Photo by M Bonnici

A 2nd Empire Shako of the 23rd infantry of the line regiment model 1852-1856. *(Courtesy of J P Morange)*

A 2nd Empire Shako of the 23rd infantry of the line regiment model 1852-1856. *(Courtesy of J P Morange)*

Inside view of the 2nd empire Shako.

Rear view of the 2nd empire Shako.

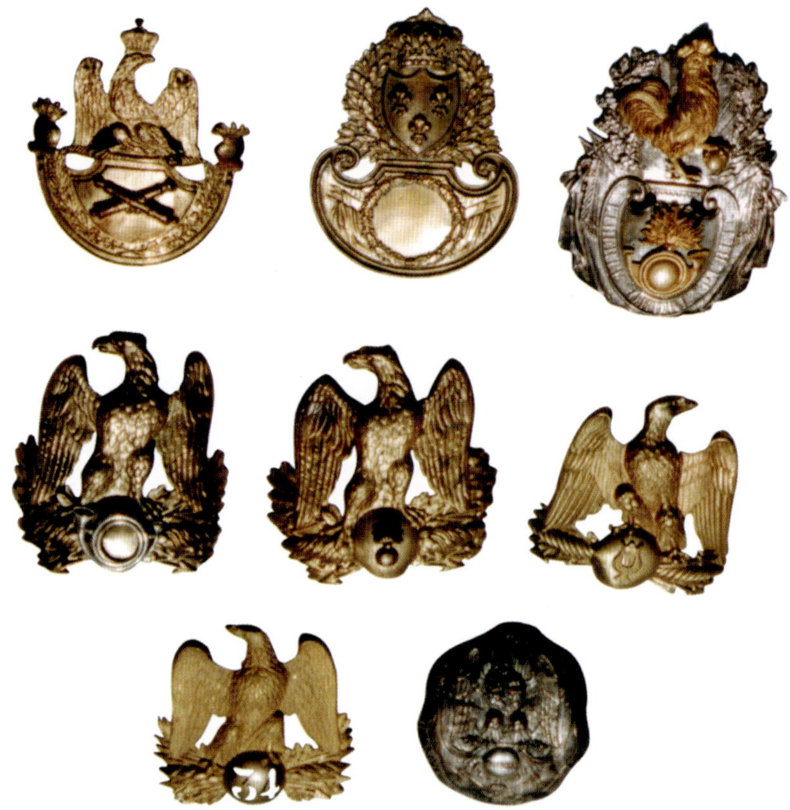

Above: A selection of Shako plates from the first empire to the second empire. *(Courtesy of J. P. Morange)* Top left: A model 1812 style plate, a replica artillery plate produced in 1840. Top centre: A model 1821 plate. Top right: An officer's model 1831 Shako plate of the voltigeurs (officer's plates were in bi-metal). Middle left: Shako plate of a voltigeur of the Garde Nationale c. 1854-1856. Middle centre: Shako plate of a Grenadier of the Garde Nationale c. 1854-1856. Middle right: Shako plate of a Musician of the infantry of the line c. 1852. Bottom left: Officer's helmet plate of the 34th Battalion of the Garde National c. 1852. Bottom right: Helmet plate of the Sapeurs Miniers de. St. Germaine c. 1852. Left: The reverse of the Shako plates.

Kepi was a 12 centimetre tall felt hat that resembled a short and soft version M1840 Shako. A rectangular peak and Hungarian style braiding was stitched onto the hat. In 1867 the popularity of this comfortable fatigue cap eclipsed the Shako, and a short Kepi that was only 8 centimetres tall took the Shako's place.

In 1873 the rectangular peak was substituted with a rounded peak.[15] A year later a stiff, higher quality version of the M1873 was issued to the cavalry. This hat was a hybrid headdress that included some elements of the late Shako. The 'Shako elements' consisted of the leather top, gold braid ribbon on the upper part of the Shako, metal fittings, and the cockade. This Shako/Kepi was heavily criticized,[16] but was abolished only in 1912. However, it was retained for parade purposes at the *Saint-Cy-Coetquidan* Military Academy, and by the *Regiment a' pied*[17] *de la Garde Republicaine*, and is still worn today.

A selection of helmet plates from the Third Republic. Top left: An Infantryman's Kepi badge model 1886-1914. Top centre: An artilleryman's Kepi badge model 1886-1914. Top right: A light infantryman's Kepi badge 1886-1914. Bottom: A Kepi badge of the chasseurs a cheval c 1886-1914.

The Shako in Great Britain

The history of the British Shako runs parallel to that of the French Shako. Whereas the first French Shako evolved from the Mirliton, the first British Shako evolved from a peculiar hat that was worn in the Imperial Austro-Hungarian Army in the 18th century known as the '*Kaskett*.' The Austro-Hungarian Model 1779 peaked *Kaskett* was a round leather hat of Pokalem form that was 16 centimetres high.[18] The first British Shako was issued in February 1800. The British 1800 pattern Shako was of a very simple form. It consisted of a closed leather cylinder fitted with a large helmet plate embossed with the Georgian coat of arms. It was slightly more than a centimetre higher[19] than the *Kaskett*, and its tall stovepipe shape earned it the name stovepipe Shako. It was found to be inadequate. Its ungainly height rendered it unsteady on the head, and it weighed too heavily on the wearer's skull. Moreover, it offered no serious protection against sword cuts. A substitute Shako was ordered as early as 1806, but was formally issued to chosen units only in 1812.[20] The 1812 pattern was referred to with various names that recalled its Napoleonic heritage, such as the 'Waterloo Shako,' the 'Belgic Shako,'[21] or the 'Wellington Shako.'[22] It was a light felt Shako with a false raised front similar to that of the 1779 Shako. The fragile felt edges were strengthened with thin leather strips stitched into the edges. It was fitted with a leather peak and stood 17 centimetres high.

Although the British troops were victorious in Belgium, they must have felt outdone by the tall '*pot de fleur*' style Shakos of their enemies, because they adopted very similar Shakos in August 1815. The Shako style known as 'flower pot type' in France became known as 'bell top' in England. The 1815 pattern British Shako was slightly higher than the Belgic Shako, but unlike the Belgic Shako it had no false front, and had the distinctive flared up shape of the French Shakos of the period. A Shako that was shorter than the M1815 by some three centimetres was sanctioned in 1835. This Shako, known as the 1835 Pattern, was reinforced with a pair of French M1806 style leather lateral strips. A French influence—that of the French M1837—was felt in the British Shako, which took the M1835's place in 1844. The 1844, or 'Albert Shako,' had the same 'trunk shape' of the French M1837, but it included some innovations. It had a back leather peak, and was only 16 centimetres tall, almost 2 centimetres shorter than its French counterpart.

The British Empire adopted another French style Shako during the Crimean War in January 1855. This time the source of inspiration was the successful French M1840-43 Shako. The British M1855 felt Shako retained the back peak of the Albert pattern, but had it not been for this feature it would have been only a shorter facsimile of the French M1840-43, standing only thirteenth centimetres high. A shorter felt Shako (around 10.5 centimetres) of an identical shape substituted this Shako in 1860. The M1860 was substituted by a still lower Shako (10 centimetres) in 1869. The latter Shako fell in disuse in 1878 when it was replaced by the blue cloth helmet of Prussian inspiration.[23]

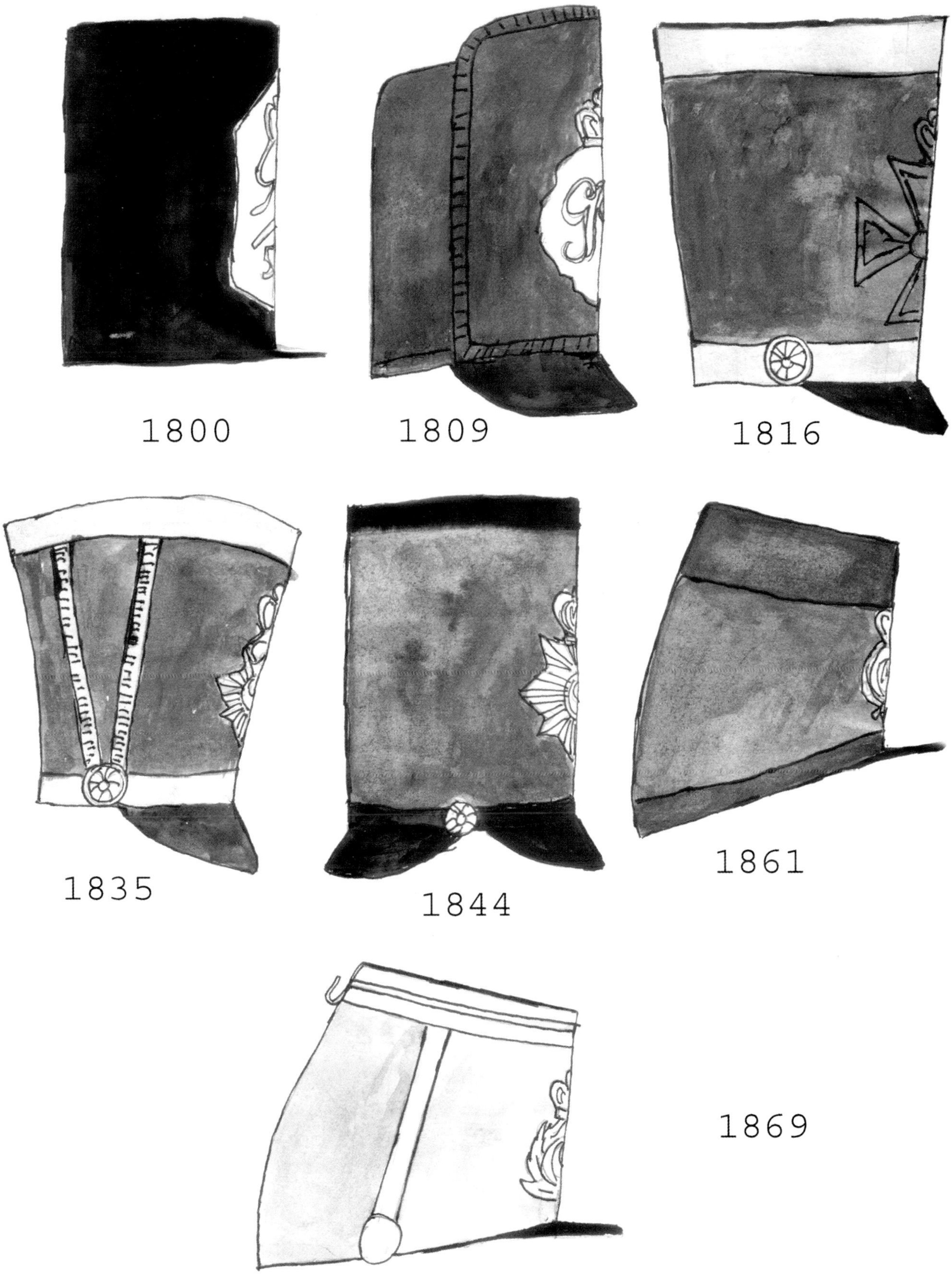

Artist's impression of the evolution of the British Shako (heights approximately to scale).

6.1.3 The Shako in the USA

The first Shakos that were worn in the USA (where they became known as tarbuckets) were based on the Shakos worn by the British Army in the Napoleonic wars, the stovepipe Shako and the 'Belgic' Shako. The first American Shakos consisted of felt versions of the British stovepipe Shako,[24] but the Americans soon switched to the use of leather Shakos. An all American Shako pattern appeared around 1814. The Americans merged the Belgic pattern and the stovepipe Shako and created a leather Shako, which incorporated a false front. Fine examples of this hat are kept at the Smithsonian Institute.[25] Around a year later the American Army adopted another leather Shako pattern that did not have a false front. The latter was fitted with a leather havelock.[26]

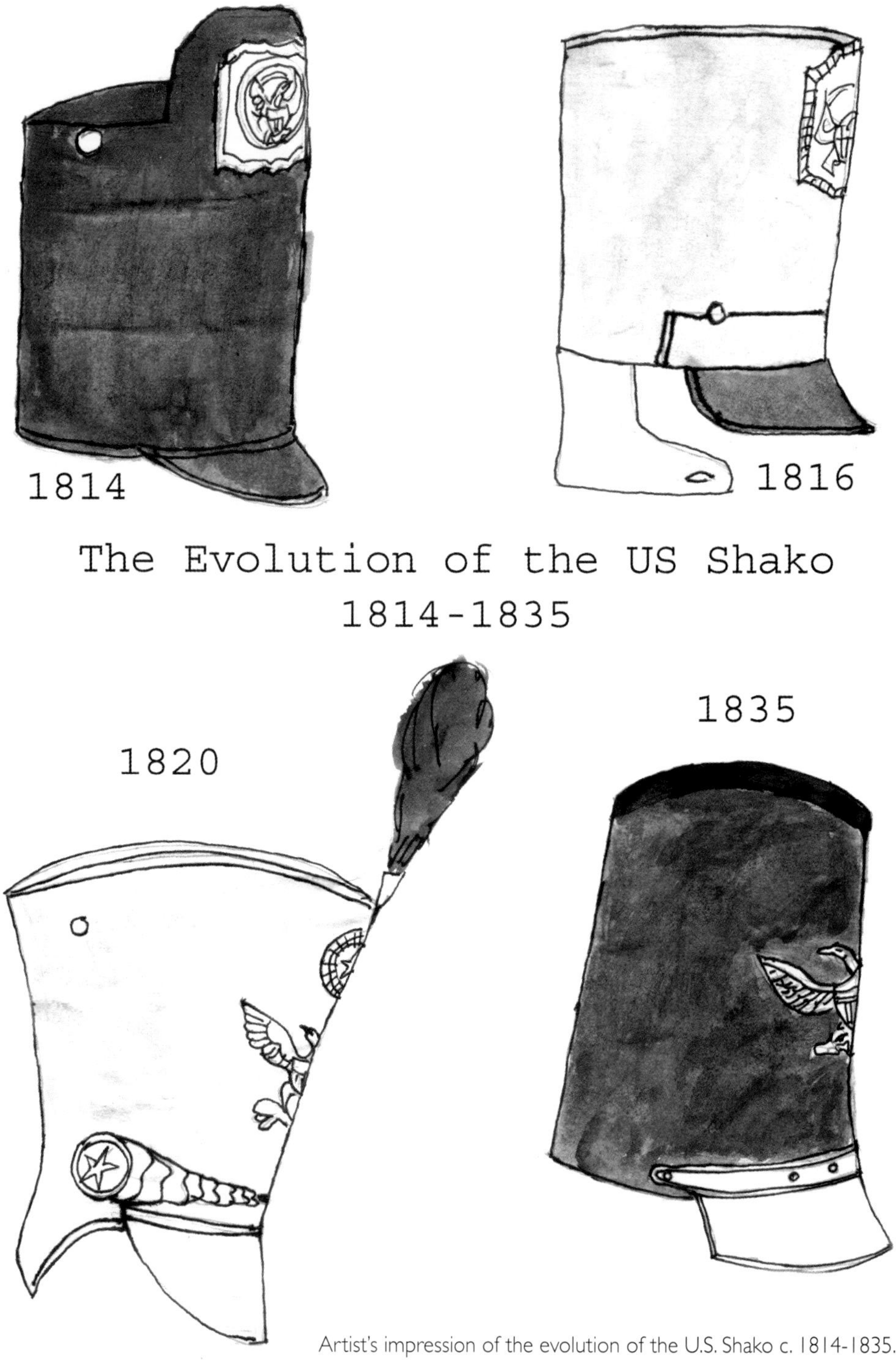

Artist's impression of the evolution of the U.S. Shako c. 1814-1835.

The influence of the French flowerpot Shakos reached the USA with the American 1820 pattern. The 1820 pattern was a tall black leather Shako of flowerpot shape fitted with a large brim and having a parallel neckpiece.[27] In 1835 another equally flamboyant Shako appeared on the American military scene. It had a tall cylindrical body with a square shaped brass, which was occasionally made of brass. The success of the French M1840-43 spread to the USA in 1851 when the U.S. army adopted a felt Shako in the '*tilted forward*' French style.[28] When the Civil War broke out French M1860 Shakos were directly imported from France by the troops of the Union[29] and fitted with Union fittings. At this time America discovered another successful French style headdress, the Kepi. In 1872 the U.S. army adopted a more squattish version of this Shako, but it was replaced with the Spike Helmet in 1881.[30]

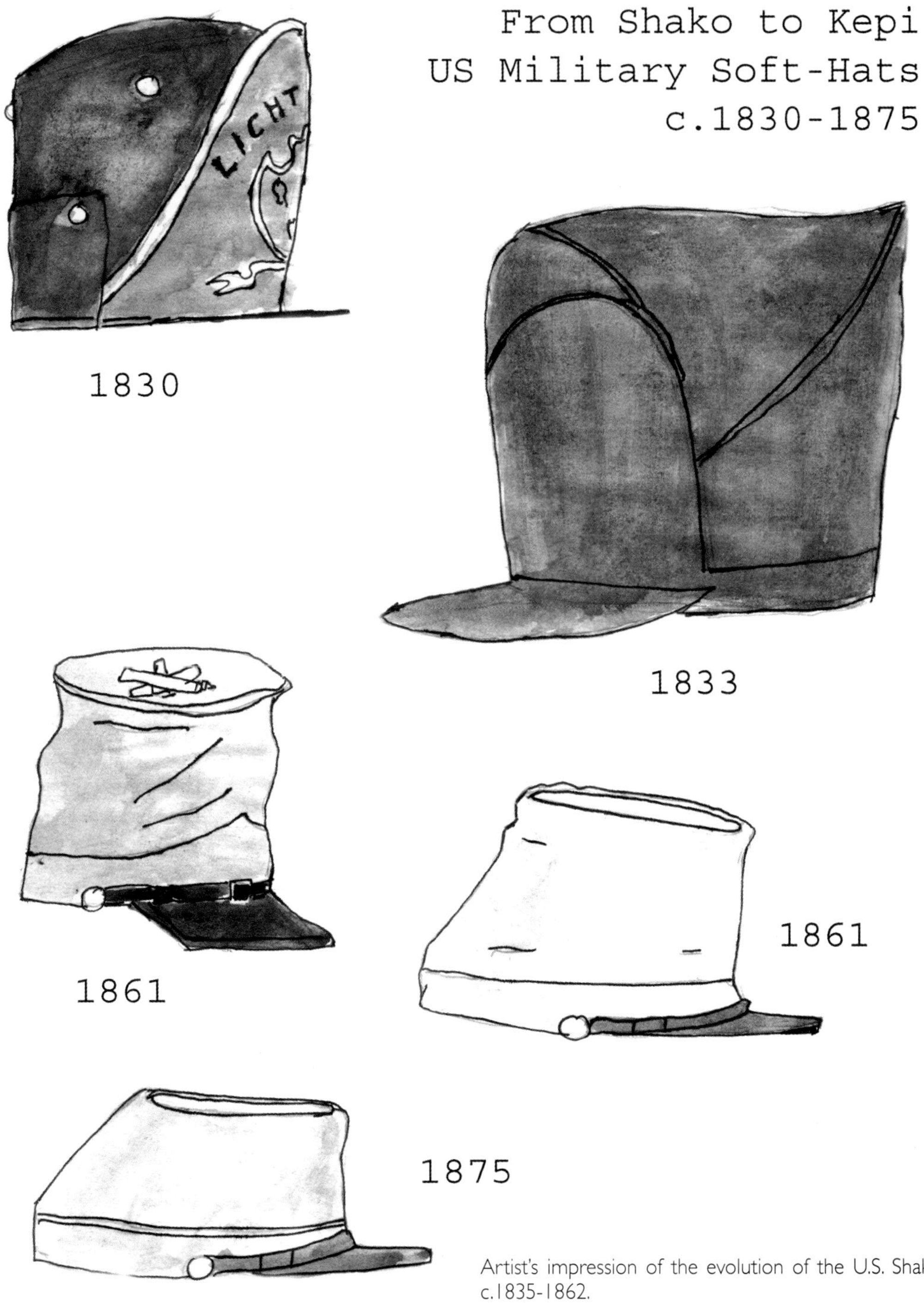

Artist's impression of the evolution of the U.S. Shako into the Kepi c.1835-1862.

A Military musician's M 1870 Shako from a Ohio militia unit. (Private collection - Malta) Photo by M Bonnici

Side view of the U.S. Shako. Photo by M Bonnici

Inner view of the Shako. The Shako was stripped of its liner, exposing the cardboard framework. Photo by M Bonnici

A late 19th century peaked cap with its original oil cloth cover. Similar oil cloth covers were used in conjunction with shapkas and Shakos in the 19th century. (Private collection - Malta) Photo by M Bonnici

The Shako in the Rest of the World

In the 19th century the uniforms and traditions of the French Army were imitated by the world's leading armies. The armies that mostly followed France's example were Belgium,[31] Switzerland,[32] and the Netherlands.

French style Shakos were also worn in Russia in the first half of the 19th century. A directive of April 8, 1843, formally decreed the use of a French M1815 *'pot de fleur'* style Shako. The use of this Shako was short lived, because it was replaced with the Picklehaube style helmet on May 9, 1844.[33] However, the Imperial Russian Army reverted to use of the Shako. In 1870 the Russians wore an 1860 style Shako.[34] In the early 20th century Russia adopted a stubby bell top Shako (Kiva) with concave top as worn by the imperial Russian army during the Napoleonic wars for parade use .[35]

The first Shako appeared as a test model in Prussia in 1801,[36] but Shakos were formally adopted only in 1807. The Prussian M1807 was quasi-identical to the French M1806.[37] The Prussians adopted another Shako based on the French M1812 in 1812, but in 1813 the Prussian army was equipped with a Russian flowerpot Shako with concave top.[38]

The Shako fell in disuse in Prussia in 1842 when the Spiked Helmet was formally approved, but a leather version with a leather back peak was re-introduced for the *Jaeger* battalions in March 1854.[39] This tall helmet was reduced in size in 1860.

Ventilation holes were added in 1888, and this Shako was fitted with Picklehaube style chinstrap mounts in 1891.

This pattern and cheap economy felt versions of this helmet remained in use up to 1916, but the police retained it for many years to come.[40] Other German states continued to make use of the Shako long after the introduction of the Picklehaube. Hamburg used a French M1840-43 Shako in the 1850s,[41] and a short felt version of the French 1860 in the 1860s.[42] Hannover used a French M1837 style Shako in the 1860s,[43] and Braunschweig a French M1860 style from 1863 to 1867.[44] Saxony was one of the states that retained its short felt Shako for the longest time, substituting it only in 1914.[45]

Initially, Denmark closely followed Prussian Shako styles. The Danish M1860 Shako was a felt and leather version with the all leather Prussian M1860 for *Lanwdehrinfanterie*,[46] but towards the late nineteenth century they employed French M1872 style Shakos for the *Garde Hussar Regiment*.[47]

Switzerland and Sweden used distinctive Shakos of Kepi form fitted with odd back peaks in the late 19th century.[48]

Artist's inpression of the evolution of the Portugese Shako.

A Prussian Landwehr Infantry Officer's Shako c. 1840. *(Courtesy of of Hermann Historica)*

Interior view of the Danish Shako. The fine red silk liner bears the hatter's name CL. Seifert. Photo by M Bonnici

Another view of the Danish Shako. Note the triple band braid. Photo by M Bonnici

6.1.5 Battles fought in Shakos

The Shako was widely used in the Napoleonic Wars. The earlier Napoleonic wars were fought in the French Model 1806, while the later wars witnessed the use of the M1812. Although the British 'Waterloo' Shako was used at the battle of Waterloo, it appears that the stovepipe Shako worn in the peninsular war was also worn in this battle.[49]

The British 1816 pattern Shako was used in the battles of the early British campaigns in India and South Africa.[50] The British 1835 bell top pattern was used in the Afghan war.[51]

The French M1815 Shako[52] and the Model 1840-43 appear to have been extensively used in France in the uprisings of 1848.[53] The Turkish,[54] French, and Italian troops wore the French M1840-43 pattern Shako in the Crimean war.[55] The same Shako was used by Oudinot's troops when the French intervened in Italy. At this time the popularity of the *Casquettes D'Afriques*[56] was widespread, and this Shako/Kepi saw service in the French campaigns in North Africa and in the Crimean war. The British used the 1844 Albert Pattern Shako in the early part of the Crimean War, but started to use a Shako of French inspiration in 1855.[57]

The Italian War of Independence appears to have been fought in French 1840-43, 1852, and 1860 pattern Shakos. The tilted forward appearance of the French M1840-43 Shako can be discerned in various illustrations of the period.[58] In some of these illustrations Shakos are covered with oilcloth covers.[59] The popularity of the French Shako was universal, and militia units in the early civil wars used 1860 pattern French Shakos.[60] The Franco-Prussian war witnessed the decline and fall of the French Shako. Following the French defeat, the armies of Europe appear to have ceased to emulate French Militaria, and started to follow the military traditions of the Prussian victors. Popularity of the Shako gave way to the popularity of the Picklehaube. This does not mean that the Shako was phased out completely. The Austrian infantry and the Prussian Jaegers still wore the Shako in the early First World War.

6.1.6 Collecting

All 19th century Shakos are rare, and are met with a fierce demand on the collectors' market. British models often sell for thousands of dollars. They are met by a great demand in Britain itself. In 2003 a British dealer offered an officer's Shako of the 15th Hussars (circa 1820) at USD 10,400,[61] and a pre-Waterloo Shako of a trooper of the 10th hussars at the same price.[62]

The quasi-unobtainable First Empire Shakos fetch similarly prohibitive prices. The latter are exceptionally rare, and the author did not locate First Empire Shakos being offered for sale by dealers in 2003. Consultations with an advanced French

A Dutch militia Shako stripped from its fittings c 1875. (Private collection - Malta) Photo by M Bonnici

A late 19th century rare French Shako box for an M 1872/95 Model Shako. (Private collection - Malta) Photo by M Bonnici

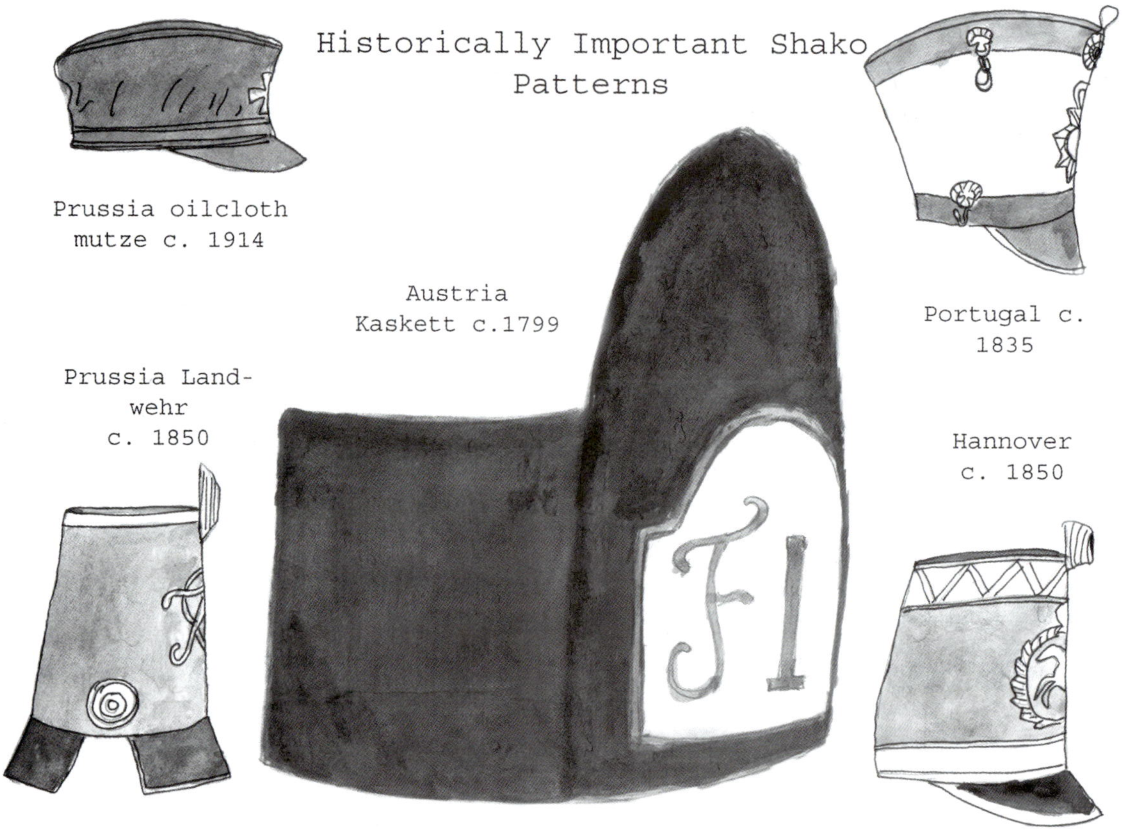

Artist's impression of a series of historically important Shakos.

collector of napoleonica[63] yielded some interesting valuations. This collector estimated an M1806 pattern trooper's Shako of the 1st Hussar Regiment at around USD 5,000-7,000. The same collector identified a Model 1810 Shako at around USD 4,000-5,000. This collector also estimated a Shako of the 4th Regiment of Napoleon's Gardes d'Honneur at USD 7,000-USD 9,000.

French Republican and Napoleonic Shakos from the 2nd Empire luckily sell for more accessible prices. A good M1840-43 Republican Shako of the Mobile Infantry was valued at USD 500-700, while M1852 patterns were valued in the region of USD 600-800. In 2003 a French dealer was offering a hard to find M1840-43 Shako of a Lieutenant Colonel at around USD 1,000.[64] The same dealer offered a rarer M1815 with Orleanist fixtures at USD 1,200.[65] Late French (1872 Model) Shakos sell in the USD 400 price range. In summer 2003 a French dealer offered a M1872 of an Adjutant of the Infantry at around USD 490.[66] The same dealer offered an M1872 Shako of an officer of the infantry at USD 350,[67] and an M1872 Shako of a *Sous officier* at around USD 500.[68]

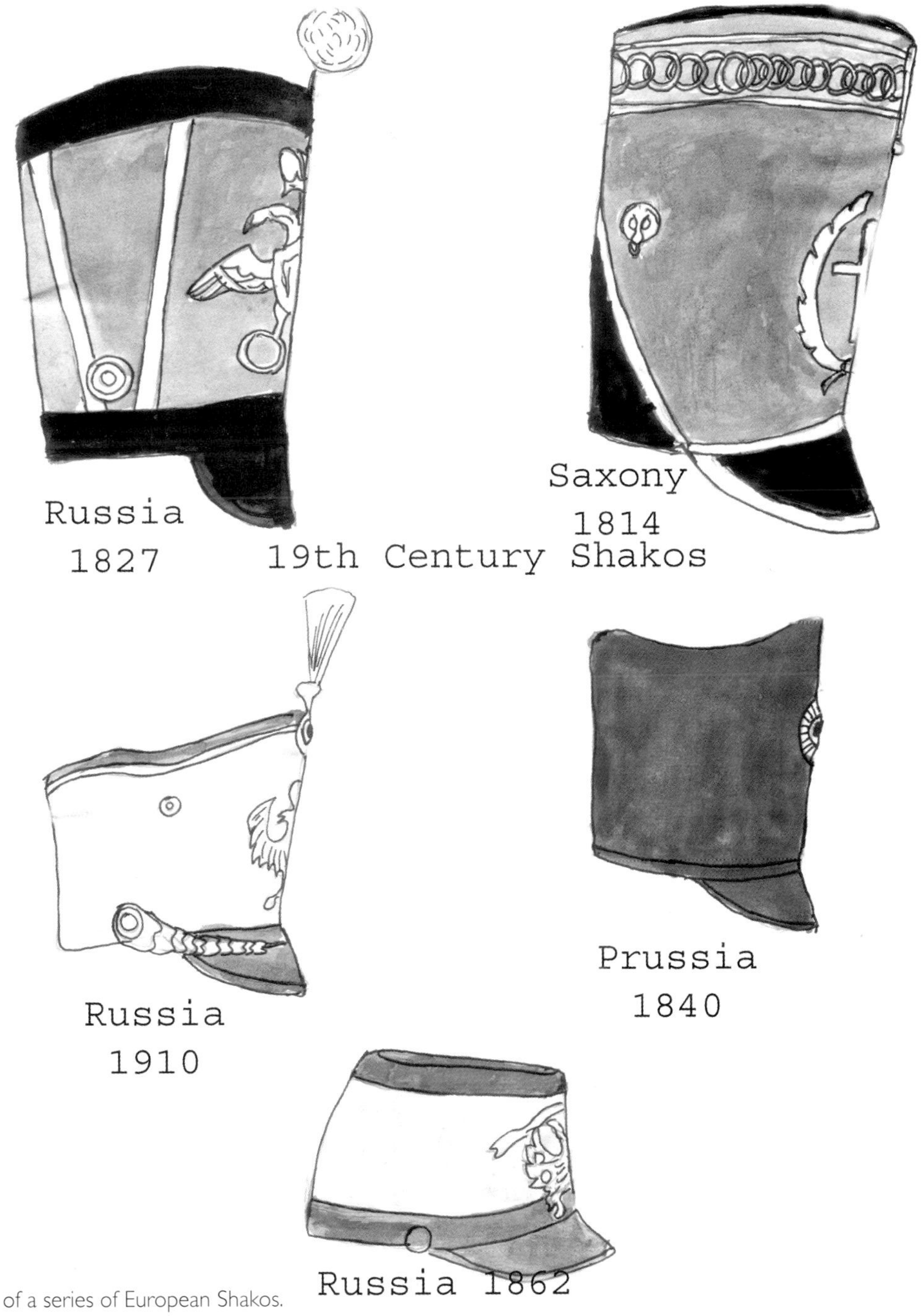

Artist's impression of a series of European Shakos.

Artist's impression of European shapkas.

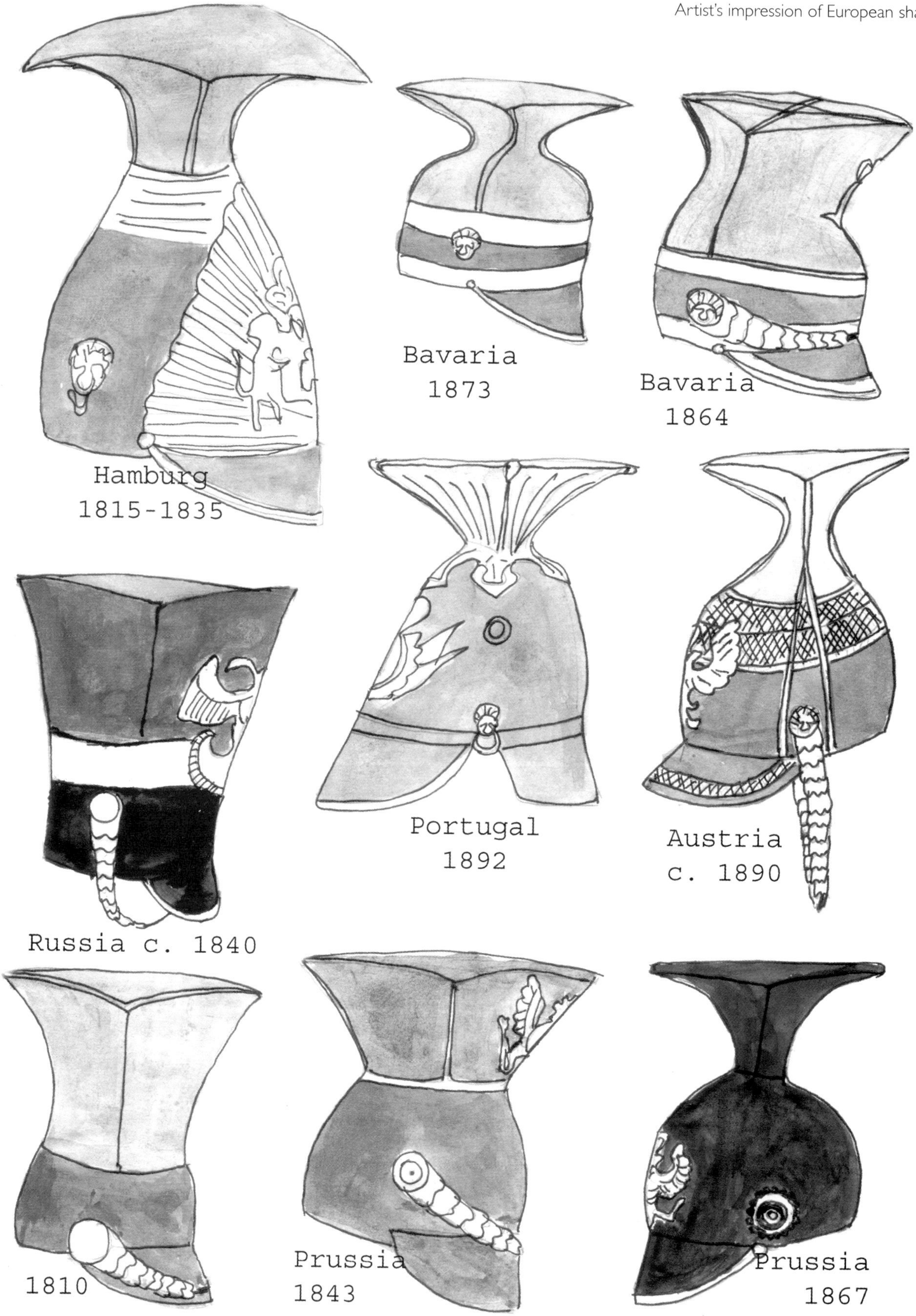

collector of napoleonica[63] yielded some interesting valuations. This collector estimated an M1806 pattern trooper's Shako of the 1st Hussar Regiment at around USD 5,000-7,000. The same collector identified a Model 1810 Shako at around USD 4,000-5,000. This collector also estimated a Shako of the 4th Regiment of Napoleon's Gardes d'Honneur at USD 7,000-USD 9,000.

French Republican and Napoleonic Shakos from the 2nd Empire luckily sell for more accessible prices. A good M1840-43 Republican Shako of the Mobile Infantry was valued at USD 500-700, while M1852 patterns were valued in the region of USD 600-800. In 2003 a French dealer was offering a hard to find M1840-43 Shako of a Lieutenant Colonel at around USD 1,000.[64] The same dealer offered a rarer M1815 with Orleanist fixtures at USD 1,200.[65] Late French (1872 Model) Shakos sell in the USD 400 price range. In summer 2003 a French dealer offered a M1872 of an Adjutant of the Infantry at around USD 490.[66] The same dealer offered an M1872 Shako of an officer of the infantry at USD 350,[67] and an M1872 Shako of a *Sous officier* at around USD 500.[68]

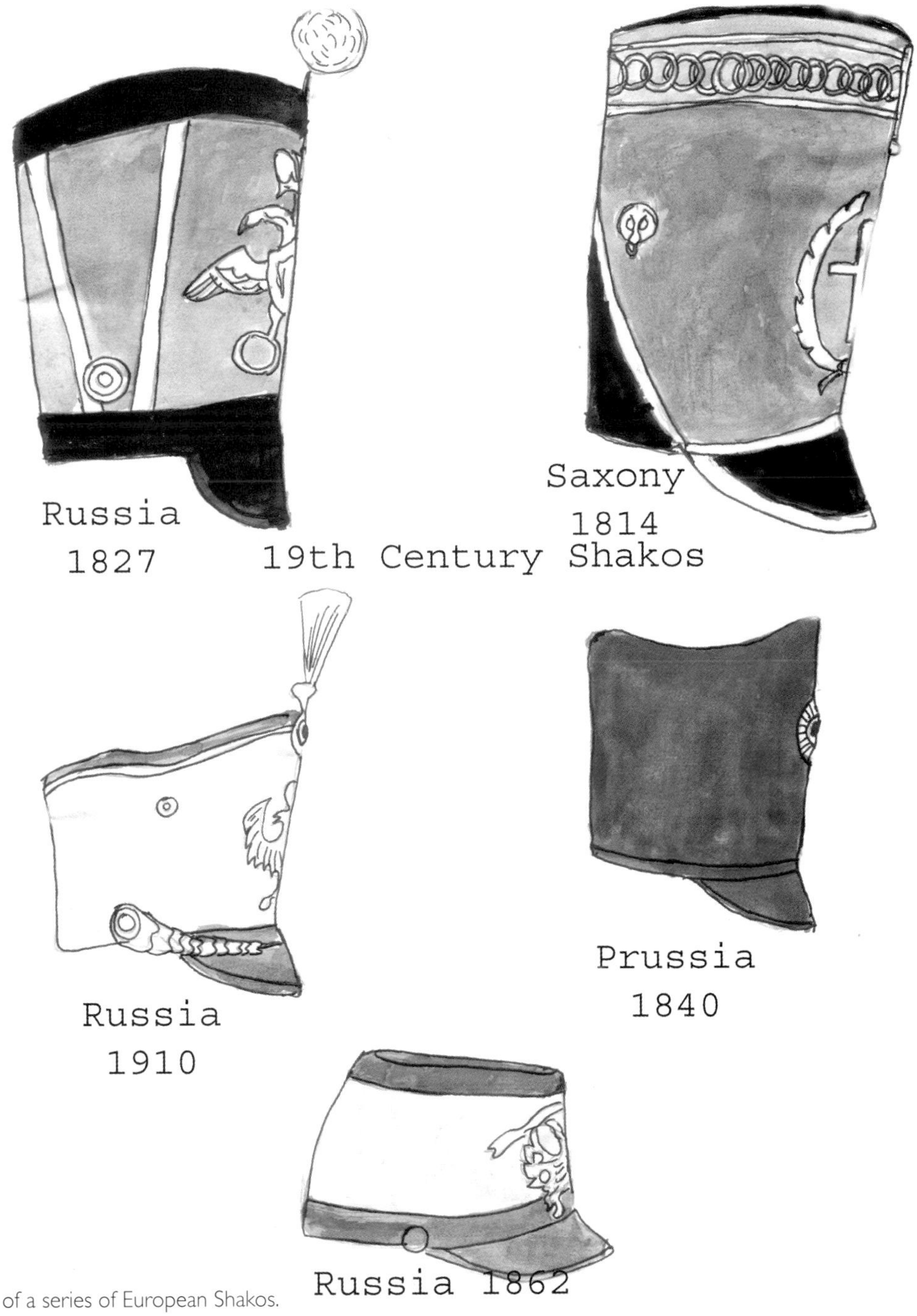

Artist's impression of a series of European Shakos.

German Shakos are also in great demand. First World War Shakos in above average condition tend to sell for over USD 400, with the rarer examples selling for significantly higher prices. A British dealer offered a Schaumburg-Lippe Shako dated 1915 at USD 948 in 2003.[69] Russian parade Shakos from the same epoch sell for significantly higher prices. A British dealer offered an Imperial Russian officer's Kiva with the distinctive concave top at USD 5,440.[70] Prices are becoming more and more prohibitive by the day, as the market is gradually drying up. Yet, budding Shako collectors should not be discouraged. Collectors with a limited budget may acquire fine examples of the 20th century Shakos worn by the minor armies of Europe at accessible prices. Dutch Shakos sell for inferior prices than those of their German equivalents, but they still pride a remarkable quality. Even the Shakos that are currently issued to military recruits make their presence felt and still sell for very reasonable prices.

Numerous fake Shakos are in circulation. Unfortunately, the oldest replica First Empire Shakos were produced as early as 1840 for the return of Napoleon's corpse to France, and in 1854 for the fiftieth anniversary of the first empire. These early replicas are indistinguishable from the originals because the very same hatters that dressed Napoleon's troops made these replicas. Replica French Napoleonic War Period Shakos appear to have been in production in late Victorian Britain, but the creators of these replicas did not follow original Napoleonic patterns as closely as their French colleagues. Excellent replicas were produced for the film industry, yet the fittings of these replicas lack the detail of the originals, while the Shako bodies were not made in the heavy quality of the originals. A large amount of scholarly replica French and British Shakos are being produced for the re-enactors' market. The author had the opportunity to observe specimens of these replicas. Some of these expensive Shakos were made from the correct type of materials (felt/cardboard and leather) and with an eye for detail (including the distinctive tooled leather peak), but in some examples the proportions of the hats and low quality of materials used is not up to the standards of the originals. Needless to say, the latter replicas do not have the aging characteristics and dating features (including tailors' labels, etc) of the originals.

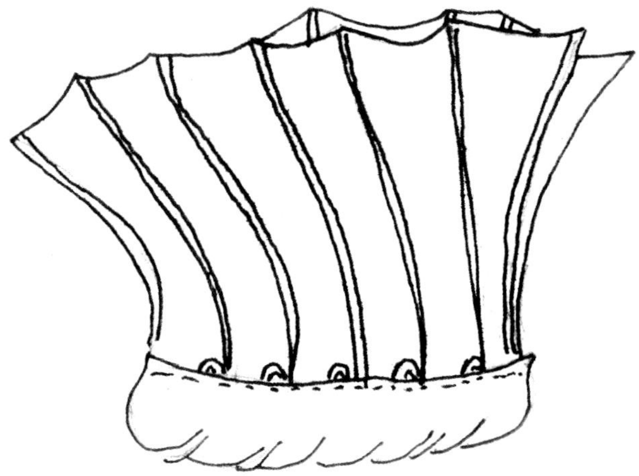

Artist's impression of a possible source of inspiration of the early shapkas.

6.2 The Shapka

"Across the hall came a man in a ragged soldier-coat and shapka, whose face was familiar; I recognised Melnichansky, whom I had known as the watch-maker George Melcher in Bayonne, New Jersey, during the great Standard Oil strike. Now, he told me, he was secretary of the Moscow Metal-Workers' Union, and a Commissar of the Military Revolutionary Committee during the fighting...."

Leo Tolstoy
Anna Karenin
1875-1877

In the author's opinion, the origins of the shapka may be traced back to a late 18th century helmet. In the early 18th century the flamboyant 'winged' Zischagge of the Polish hussars evolved into an odd helmet that consisted of a shallow

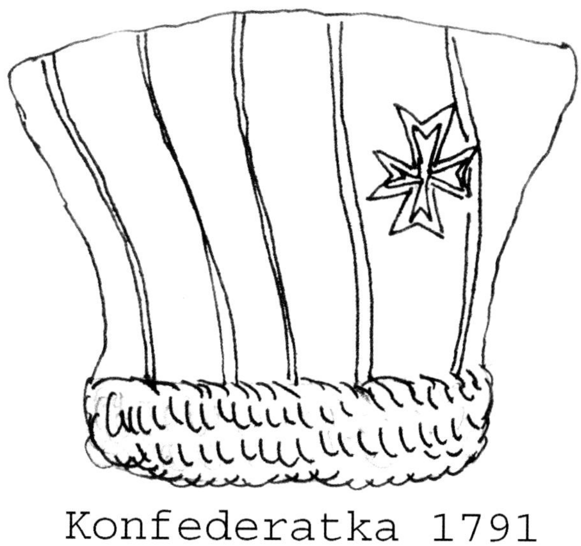

Artist's impression of a possible source of inspiration of the early shapkas.

hemispherical bowl constructed from overlapping metal plates. The bowl of this helmet was fitted with a Zischagge style nasal guard and a pair of large metal radiating 'wings' on the sides.[71]

It appears that by the mid-18th century cloth versions of this hat were in production. These hats consisted of a fur trimmed crown and five panels of cloth that formed a square over the wearer's head.

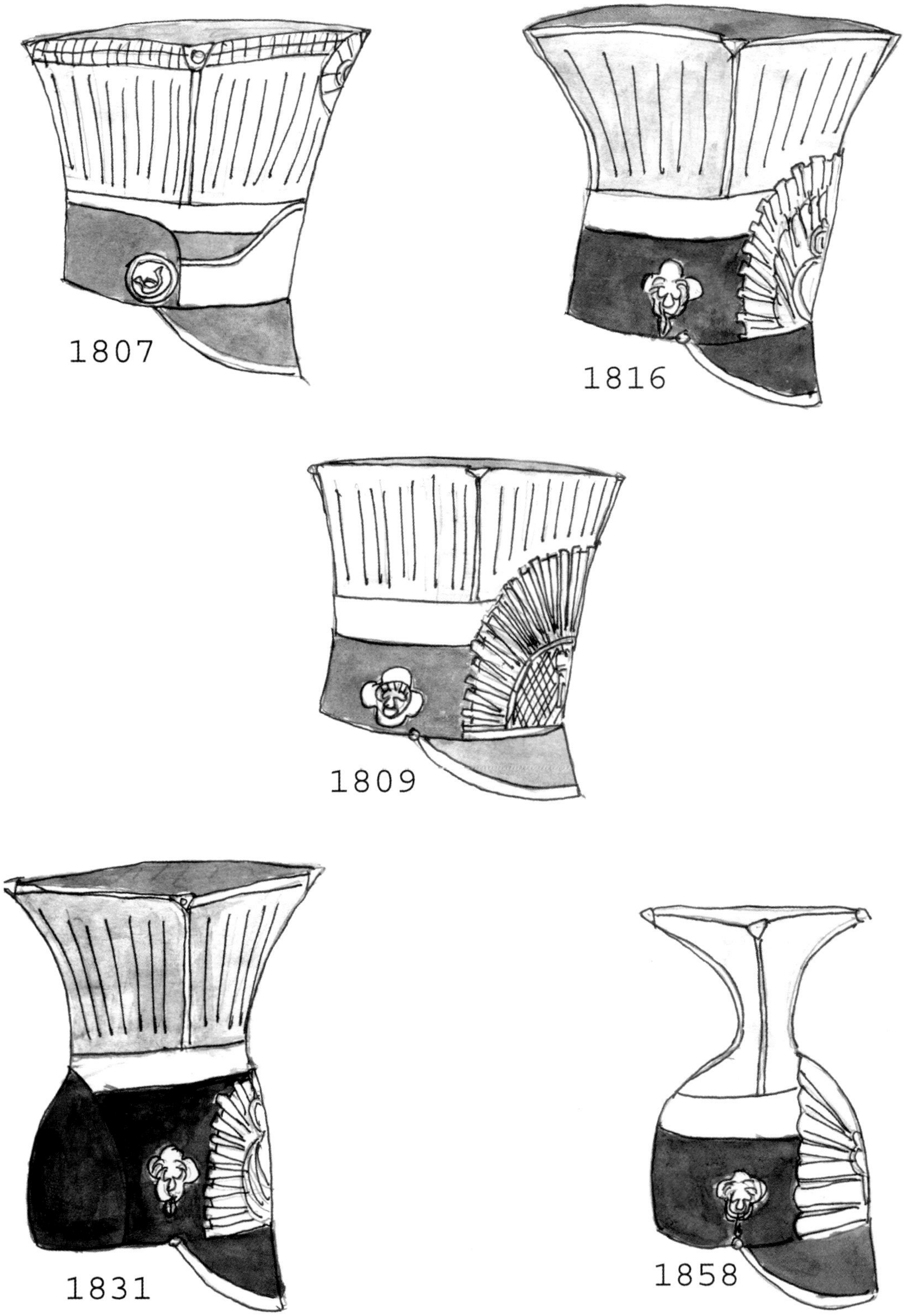

Artist's impression of the evolution of the French shapka.

Artist's impression of European shapkas.

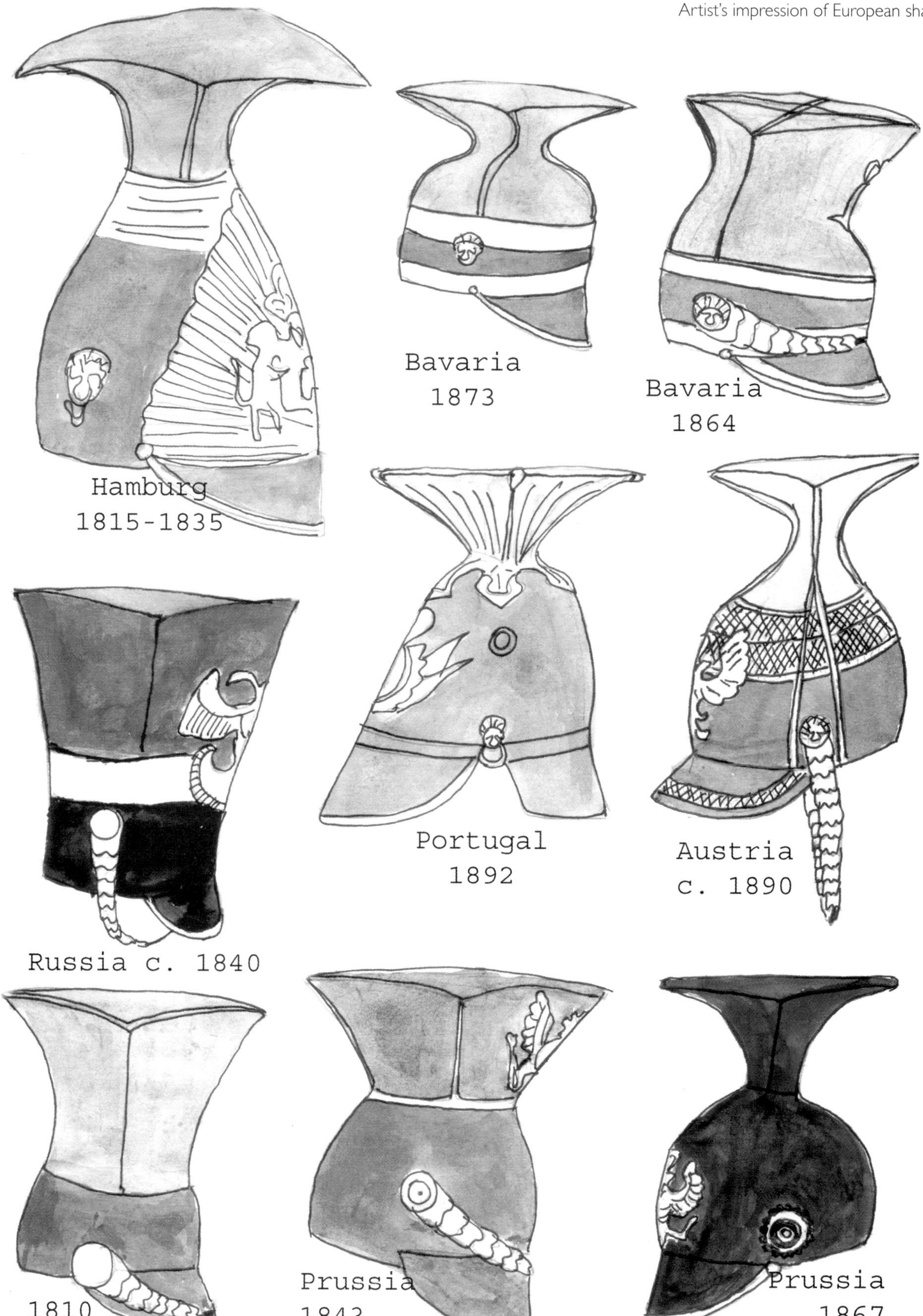

6.2.1 From France to the USA

France introduced this hat pattern in 1799 when the legendary *Legion Polonaise* (later re-named *Legion D'Italie)* was equipped with the square shaped hat, known as a Konfederatka. Eight years later Napoleon equipped the lancers of the *Legion Polacca-Italienne*, the *Lanciers Polonaise de la Garde*, and the *Legion de la Vistule* with the peculiar headgear.[72] However, it appears that shapka patterns were not standardised. Officers seem to have designed their own shapkas, while troopers wore cheaper versions that consisted of a shallow black leather skull stitched onto a brass bound peak and a permanently lifted rear peak, surmounted with a rigid scarlet cloth square top. The shapka was reinforced by a large helmet plate, a pair of large chinstrap bosses fashioned in the shape of lions' heads, and a large cockade. This shapka fell out of use with the fall of Napoleon.

In 1816 a similar shapka was issued to the *Garde Royale*, but this shapka was slightly taller than its Napoleonic predecessor and narrowed towards the waist. A slightly taller (11.5 centimetres) helmet with a thinner 'waist' was introduced in 1845. Another shapka of the same height with a narrower waist and a shaped leather skull was introduced in 1857.[73] When the lancers were abolished in 1870 the shapka fell in disuse in France.[74] Similar shapkas to those worn in France were worn in Belgium[75] and Imperial Russia.

Artist's impression of the standard Other Rank's French Napoleonic Shapka.

The success of Napoleon's bold polish lancers solicited the emulation of most of his foes. Although the origins of the British Lance Cap may trace back to the Konfederatka of the Uhlans Britanique in 1793,[76] it was the Napoleonic experience that granted impetus to the use of the shapka in Great Britain.

Above: A pair of Frech shapka plates. Left: A plate from the Louis Philippe Era (Garde nationale a cheval de Province) (c. 1837). *(Courtesy of J. P. Morange)*. Right: A Second Empire plate (Guarde nationale a cheval de Paris) (c. 1852). Below: The reverse of the Czapka plates.

A model 1810 shapka of the 2nd regiment de Chevau - Legers Lanciers of the Imperial Guarde (reconstructed c. 1840-1860). *(Courtesy of J. P. Morange)*

Side view of the reconstructed shapka—the fittings are all original first empire period.

The same helmet from another perspective. Although it is a very old replica, it is virtually indistinguishable from an original, as it was reconstructed from components of an original First Empire cap around 1840.

Rear view of the shapka.

According to Carman,[77] the first British czapka (known as a Lance cap) was issued in 1816. It was copied from the French 'trencher cap,' false back peak and all, but appears to have been taller and heavier than its French counterpart. It was also characterised by a substantial leather peak.[78] The British Lance cap also had a more 'modern' appearance than the box shaped helmet of the *Lanciers Polonaise*, because the British M1816 had a much more slender waist than the French shapka.[79] Rowe[80] recently published the following extract from the 1822 Dress Regulations on this early Lance cap:

"The square top was ten inches square, and it was kept firm by a wire skeleton that held panels of cloth that were reinforced with quilted cane strips—a manner of construction that was retained in later specimens."

The days of the 1816 shapka were counted when a Horse Guards circular announced new dress regulations in 1828. A new shapka was formally issued in 1831. The new shapka was approximately seven centimetres shorter than the 1816 pattern, and it also had wider sides. The 1831 regulations also decreed a new range of Lance cap plates. At this point the British Lance cap was still fitted with the back peak. This important dating feature was abolished with the changes of 1856.

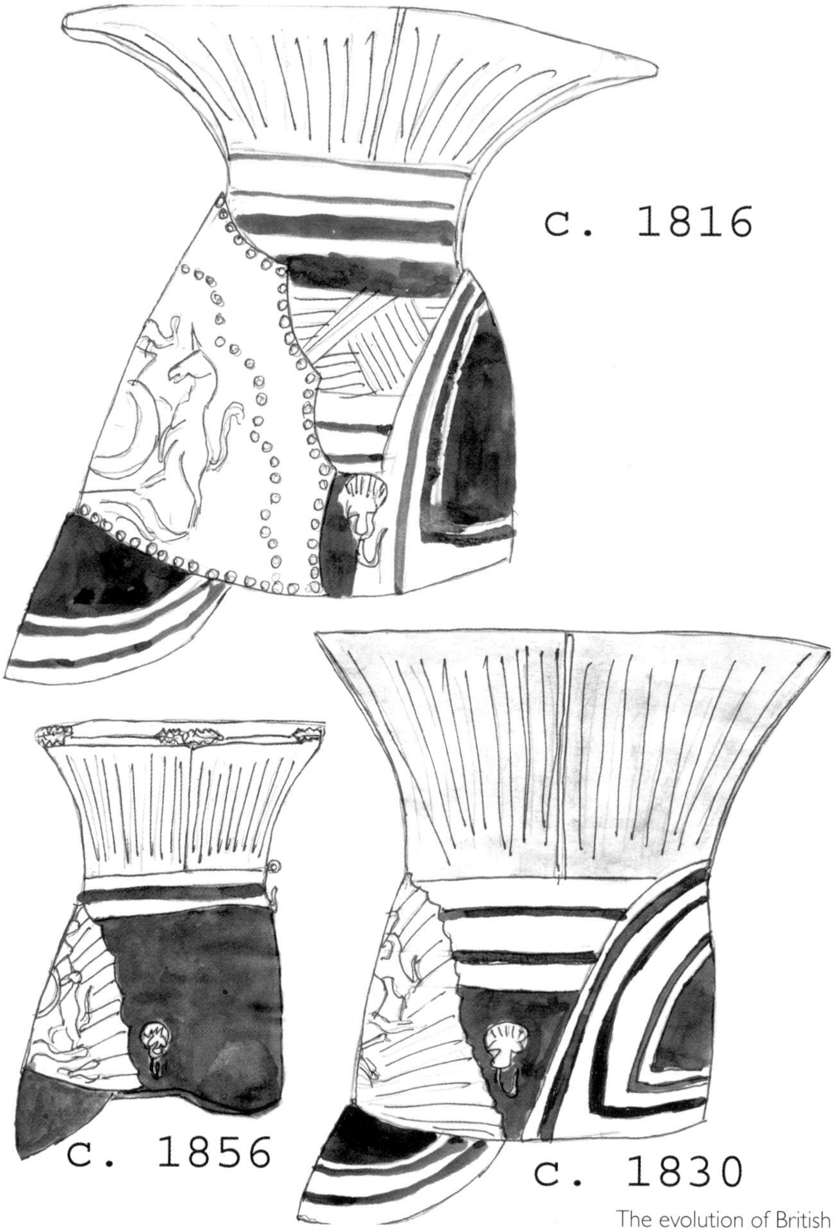

The evolution of British shapka 1816-1900.

A relic post 1856 pattern trooper's shapka of the 17th Lancers. This Shapka was published in Treasures of Malta Vol VI No 3. (Private collection - Malta)†Photo by M Bonnici

Front view of shapka; the plate was acquired separately from the shapka. The plate dates c. 1880. Photo by M Bonnici

The shapka with its plate removed. Photo by M Bonnici

Rear view of the shapka. The absence of a back flap indicates that the helmet is post 1856.Photo by M Bonnici

Detail of the famous skull and cross bones motif of the Seventeenth Lancers. Photo by M Bonnici

A new shapka pattern was introduced in 1856. The 1856 pattern was a shorter helmet consisting of a globosely moulded leather cap fitting closely round the wearer's head. The trencher top was also shorter. Different shapka patterns were prescribed for the officers and troopers. The officer's cap was embellished with straps of gold braiding around the peak, the bottom of the cap, and the waist. The trooper's version had no braiding and a less globular cap. Paradoxically, the officer's helmet was shorter than the trooper's helmet, but both shapkas were in the French Shako style, having taller backs and lower fronts, making the hats look as though they were titled forwards. The officer's helmet was 21.5 centimetres tall from the front and 25 centimetres from the back, while the trooper's example was around 22 centimetres tall from the front and 22.9 centimetres from the back.[81] Initially, the Lance cap top was made of leather, though it was later made of cloth.[82] This shapka was worn on the battlefield up to early in the First World War. Rowe published a rare photo taken in 1914 that shows British lancer troops wearing 1856 pattern shapkas in marching order.[83] After the mid-First World War the shapka was relegated to a purely parade headgear and disappeared from the military scene by the Second World War. Fibreglass shapkas were issued to bandsmen in the 1980s, and it appears that this unit is in the process of being issued with replica 1856 pattern shapkas[84] made from original materials.

The shapka appeared in the German states during the Napoleonic wars. The earliest German shapka illustrated in Muller[85] dates back to 1813. It is a tall saxon facsimile of the French Napoleonic shapka, but the saxon shapka appears to have had a narrow waist. The same author illustrates a quasi-identical Danish shapka from the same period; this shapka has a leather back brim reinforced by a brass trim.

According to Johansson, the Prussian Uhlans adopted the shapka as late as December 1843.[86] This shapka consisted of a leather body surmounted by a tall square top. In the Prussian shapka the helmet plate was fitted on the front panel of the shapka, rather than the leather helmet body.[87] Similar shapkas were worn in Imperial Austria,[88] Militia units in the United States,[89] and Poland.[90] In 1862 the 1843 pattern was slightly modified, with the addition of a brass peak trim on troopers' patterns and a silver trim on officers' patterns. This shapka was replaced by a shorter all-leather shapka in 1867. The size of the mortarboard was further reduced in 1894. This shapka was worn well into the First World War, when ersatz (economy issue) felt and metal patterns of this shapka were issued to the troops. The shapka was abolished from the Prussian army in 1917.[91]

The evolution of the Portuguese shapka reflects a heavy British and Prussian influence. It appears that the Portuguese adopted it in 1835 when the *Lanceiros De Rainha* wore a facsimile of the British 1816 pattern shapka.[92] By 1848 the

A First World War period photo postcard that illustrates German Uhlans wearing shapkas on the battlefield.. (Private collection - Malta) Photo by M Bonnici

Portuguese had even plagiarised the skull and cross bones motif of the helmet plate of Britain's 17th lancers. A Portuguese officer's 1852 pattern illustrated by Soares Branco[93] is quasi identical to a British shapka illustrated by Rowe.[94] However, it should be pointed out that the shapka worn in Portugal also possessed characteristics of the Prussian 1843 pattern, such as the short top and Picklehaube shaped body. In 1892 the Portuguese shapka was strengthened by a cross piece beneath the mortarboard and a brass reinforced neck flap. This shapka remained in use up to 1910.[95]

Battles fought in Shapkas

Period paintings of the battle of Waterloo depict a line of French lancers wearing tall, scarlet-topped Lance caps, thus confirming the use of the shapka at this famous battle.[96] Similar shapkas were worn in an equally famous battle—the charge of the light brigade, when the legendary 17th lancers charged the Russian guns at Balaclava in the Crimean war. At this time officers used to wear special lightweight shapkas covered in oilcloth, while troopers used to cover their shapkas with inexpensive oilcloth covers during active service.[97] It also appears that British lancers wore the shapka in the first year of the First World War.[98]. The Prussian Uhlans made extensive use of the shapka in the early First World War, and photos of uhlans wearing their shapkas in the trenches abound. The Prussian shapka was so popular in the early First World War that special *ersatz* economy issue felt and tin shapkas were issued in the early First World War.

Artist's impression of a pair of unusual shapkas.

A German reservist's Mutze c. 1915. The Mutze (fatigue cap) was derived from the *konfederatke*. (Private collection - Malta) Photo by M Bonnici

Two 1856 pattern British Shapkas being offered for sale at the MAX show. *(Courtesy of of Tom Berndt, Militaria International Magazine)*

6.2.3 Collecting

The shapka is one of the rarest and most expensive forms of military headgear. It was issued in much smaller numbers than the Shako, and is much less durable than the metal neo-classical helmet. Nineteenth century shapkas are becoming more and more expensive, and even the easier to obtain Prussian First World War examples are becoming very expensive.

In 2003 a London dealer offered for sale a late Victorian lancer officer's shapka of the 17th lancers at USD 9,120.[99] Prussian Uhlan helmets are cheaper. Another London dealer offered a Prussian Erstz Uhlan tin (trooper's) shapka at USD 1,691, and an early First World War Uhlan leather (trooper's) shapka of a Guards regiment at USD 2,103.[100] The same dealer offered a trooper's shapka of the Uhlan Regiment 1-3 at USD 2,145[101] and a 1915 dated shapka cover at USD 371.25.[102] Even the shapkas of other countries sell within the same high price range. A good quality Austrian officer's shapka was valued at around USD 2,500, and a trooper's Austrian shapka was valued at around USD 1,000 in 1998.[103]

Early shapkas are exceptionally rare. Pre-1856 British shapkas are almost unobtainable and very rarely found outside museums. In 1990 a pre-1856 other rank's cap of the 16th lancers sold at around USD 2,000.[104] French First Empire shapkas are even more expensive. A shapka of the *Chevau-Legers Lanciers* of the Imperial Guard made for the return of Napoleon's corpse to France in 1840 was valued at USD 3,000. French shapkas produced during the First Empire would sell for significantly higher prices, in the region of USD 10,000. Good quality replica shapkas are hard to find, however, prospective collectors should keep in mind that very good quality French First Empire style parade shapkas are produced for the Polish cavalry military brand, and that only a detailed knowledge of the characteristics of original helmets may save a collector some surprises. Rowe also reports that good quality 1856 style pattern British shapkas are in the making.[105]

The Busby

"The non-commissioned officers and soldiers have a false boot of black leather on the lower part of the pantaloons. The cap is of bearskin with a 'flame' (flap) of madder colored cloth. The cap is surmounted by an aigrette of white horsehair."

Gleason's Pictorial Drawing Room Companion
1848

The Shako and the shapka are not the only forms of headgear Western Europe copied from the east. The history of European headgear also owes another type of headgear to the dauntless cavalrymen of the east. In the 19th century Napoleon's hussars wore a tall cylindrical fur cap in the manner of the ferocious light Hungarian cavalry, known as a Colback. This form of headgear was previously adopted by the Prussians and Austrians and was, subsequent to the Napoleonic wars, copied by the rest of Imperial France's foes. This form of headgear was retained in use throughout the 19th century. It was worn on the battlefield in the early First World War, but it was abolished from active service by 1918. However, it remains in use for parade wear up to the present day.

6.3.1 From Colback to Busby

"Portant le noir Colback ou le casque poli,
Tous, ceux de Friedland et ceux de Rivoli...."

Victor-Marie Hugo (1802–1885)
L'Expiation

It seems the first Colbacks appeared in western Europe in the first half of the 18th century on the heads of the Prussian and Austrian hussars. The Colback was a very similar hat to the Haiduk and the Talpack. It was made in the tall cylindrical shape of the Colback and featured the flap or fly of the Mirliton. In Prussia it was known as the *Pelzmutze*, and appears to have been the distinctive headgear of the *Husaren Regiments*.[106] The Prussian Pelmutze consisted of a very tall, cylindrical brown fur hat (probably otter skin) surmounted by a long scarlet fly or busby bag. A similar hat was worn by the French *Hussards de la Liberte'* in 1793, and by the *Bercheny-Hussards* in 1785, but the popularity of this hat flowered in the Napoleonic wars. It was adopted by Napoleon's light cavalry in 1800 when the *Chasseurs et Artilleurs a Cheval De La garde Imperiale* were equipped with a black Colback with a scarlet bag. In 1803 a

A 19th century etching that illustrates French hussars around 1840 wearing Colbacks. (Private collection - Malta) Photo by M Bonnici

Front view of a Victorian M 1887 Busby. (probably Royal Horse Artillery). (Private collection - Malta) Photo by M Bonnici

Side view of theBusby. Note the 'French' style†'tilted forward appearance.'Photo by M Bonnici

did not have the right to exclusive use over the Busby in Britain, and the Busby was also worn by some yeomanry regiments and by the Royal Horse Artillery.[110]

Needless to say, the Busby was also worn outside of Britain and France. Prussian style busbies were worn in the Netherlands and Belgium. Imperial Russia used short busbies in groomed lambs wool.[111] Similar busbies were worn and are still worn in Eastern Europe.

The origins of the Prussian Busby (known in Prussia as a pelzmutze) date back to the mid-19th century. The Model 1850 pelzmutze was a very similar brown fur cap to the 'monstrous muff' worn in Victorian Britain. It was made of brown otter fur for officers, and brown seal skin for enlisted men, and was fitted with a pair of Picklehaube style convex chin-scales[112] and a leather liner. A fresh pelzmutze was introduced eight years later. The shape and design of this fur cap was not very different

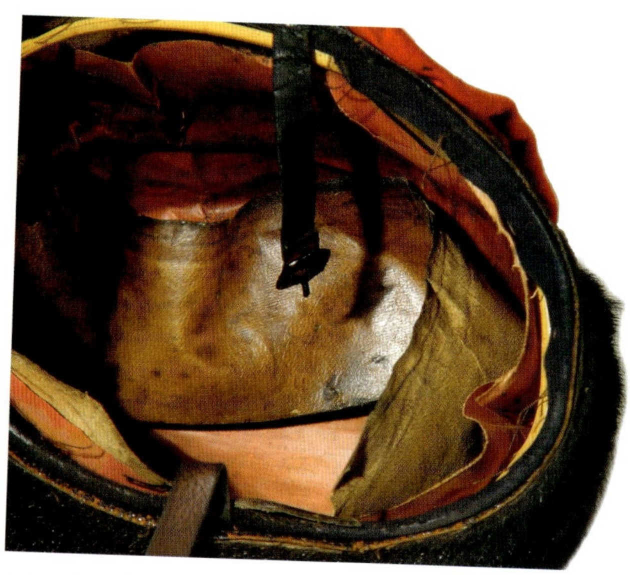

Inside view of theBusby. The lining is of red silk (faded) reinforced by a wooden strip. Photo by M Bonnici

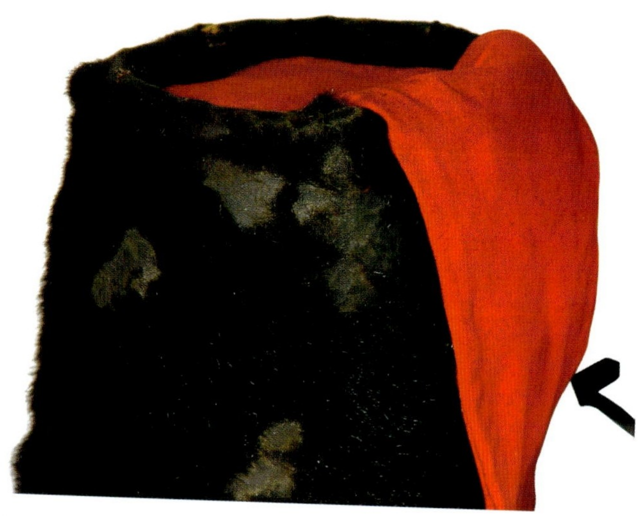

Close up view of the Busby; it was stripped from all its fittings. Photo by M Bonnici

from the earlier model, but it was slightly shorter than its earlier counterpart. Further innovations were made in 1860 when the Prussian tradition of affixing a metal bandeau to the pelmutze was born. A new pelzmutze was issued in 1866, which was the pelzmutze worn up to the First World War. It was a significantly shorter hat (13 centimetres shorter) than its earlier counterparts. The officer's version was made of brown otter fur, while the fur cap worn by enlisted men was made of black sealskin. In 1912 officers were allowed to wear cheaper opossum fur caps. The outbreak of the First World War gave rise to further changes. Brass fittings were replaced with grey pewter fittings, and a shortage of furs led the war industry to produce cheap pelmutzes made of cork, felt, cloth, and low-grade furs.[113]

A Prussian First World War period pelmutze pattern, 1915. *(Courtesy of Johan Somers).*

A Russian lambswool cavalry Busby c. 1900. *(Courtesy of Hermann Historica)*

Close-up of the maker's label on the Papapkha's liner. *(Courtesy of Hermann Historica)*

A Russian Cavalry Busby c.1900. *(Courtesy of Hermann Historica)*

Two unusual items of Belgian WWI military headgear. Left: An oilcloth Busby. Right: An artillery Busby. *(Courtesy of Johan Sommers)*

Battles fought in Busbies

The Busby was worn for over one hundred years, and this long lifespan ensured that this hat was worn in some of the most important battlefields of the 19th and early 20th centuries. An early representation of the battle of Tyn of 1744[114] confirms the use of the Colback in the 18th century,[115] in the age of the flugelmutze. The Busby was extensively used in the Napoleonic wars by Napoleon's hussars. The images of Napoleon's brave cavalrymen donning fur Colbacks was captured in some period paintings by Goya and the famous portait of La Salle by Gericault. French,[116] Russian,[117] and British[118] hussars wore the Colback in the Crimean war. British hussars wore the Busby during the charge of the light brigade. Extensive use of the pelmutze was made in Germany's wars of unification. A detailed engraving[119] from the period illustrates a Blucher Hussar wearing a tall pelmutze with a death's head helmet plate, a Westphalian hussar wearing a pelmutze fitted with a bandeau, a hussar of the guard artillery wearing a starburst badge, and an officer of the landwehr cavalry wearing a mirliton. The pelmutze was the Prussian hussar's headgear during the Franco-Prussian war. The Prussians also used it in the first years of the First World War.[120]

Collecting

The Colback represents a challenging area of collecting, because the knowledge of the average collector on the subject is rather low, and the market offers loads of opportunities. Some excellent 19th century busbies tend to pass unnoticed. The author recalls noticing an early British Busby, probably a Napoleonic war period example, selling for as little as USD 350—a ridiculous price to pay for such a historic example. It is important to keep in mind that the general rule is the higher the Busby, the older it is. Another important dating feature is the colour of the fur, as earlier examples tend to be of a brown colour, and later examples are black. French Napoleonic busbies represent the most interesting area of collecting, but unfortunately it is extremely difficult to distinguish late empire Colbacks from Louis Phillipe period and early second empire Colbacks. Moreover, French Napoleonic Busbies were frequently heavily restored (decaying black bearskin was replaced with rabbit fur), so knowledge of the furs used in the construction of originals is quintessential. Eastern European Colbacks offer another interesting area of collecting, but collectors should be extremely careful. Fine lambswool Colbacks of a Napoleonic war pattern and late pelmutze style

A much sought after Model 1894 pelmutze of the Prussian leib hussars. *(Courtesy of Johan Somers)*

A 19th century etching illustrating Russian Cavalrymen wearing busbies.

shape are still in production in Hungary and Rumania, and only a person who had the opportunity to view these modern fine hats can distinguish them from the older examples. Needless to say, Prussian pelmutzes are extremely popular among collectors, and most of the pelmutzes that appear for sale are First World War examples. In 2003 a British dealer[121] offered a pelmutze of the Prussian hussar regiment no. 16 at USD 1,867. Quality Victorian busbies sell within the same price range. A British dealer offered a Victorian Yeomanry Busby at USD 1,360 in 2003,[122] while another London dealer offered a documented Victorian 11th Hussars Busby at USD 2,227.[123] Imperial Russian busbies tend to sell within the same price range. Traces of loss of fur tend to detract from the appearance of these old fur hats, but loss of fur is an important dating feature on these hats.

Notes:

[1] Tavard op.cit p.164-165;
[2] Tavard op.cit p.166;
[3] Some infantry units were equipped with the Shako as early as 1804;
[4] Tavard op.cit pp.167-168;
[5] Wilkinson-Latham p.57;
[6] Tavard op.cit pp.173-174;
[7] Wilkinson-Latham op.cit p.57;
[8] Tavard op.cit p.179;
[9] Wilkinson-Latham op.cit p.48;
[10] Wilkinson-Latham op.cit p.48;
[11] Tavard op.cit pp.185-192;
[12] Wright op.cit p.93;
[13] Tavard op.cit pp.185-195;
[14] Rankin op.cit p.37;
[15] Tavard op.cit pp.212-219;
[16] Tavard op.cit pp.196-200
[17] The cavalry regiments of the republican guard wear M1874 style metal cuirassier helmets;
[18] Hollins & Burn: op.cit p.59;
[19] wysiiwyg;//107http://www.heroica.org/Shako.html;
[20] Wilkinson Latham p.47;
[21] Rankin op.cit p.
[22] wysiiwyg;//107http://www.heroica.org/Shako.html;;
[23] Wilkinson-Latham op.cit pp.48-49;
[24] Wilkinson-Latham op.cit p.71;

[25] Waverley Lewis op.cit fig.16;
[26] Waverly Lewis op.cit figs19-20;
[27] Wilkinson-Latham figs. 28-32;
[28] Waverly Lewis op.cit fig. 48;
[29] Waverly Lewis op.cit fig.57;
[30] Wilkinson-Latham op.cit p.74;
[31] The author noticed a Belgian facsimile of the French M1840-43 and the French M1874;
[32] The Swiss Army used an M1837 in the 1840's, a French 1840-43 style Shako (with a heavier lether top) in 1855 and an M1860 style in the1860's vide Rankin op.cit pp.34-39'
[33] Vikovatov op.cit;
[34] Rankin op.cit fig. 64;
[35] Rankin op.cit fig 80;
[36] Kube op.cit p.37;
[37] Muller op.cit fig.184;
[38] Muller op.cit fig.185;
[39] Johanssen op.cit p.73;
[40] Johanssen op.cit p.74;
[41] Muller op.cit fig.211;
[42] Muller op.cit fig.217;
[43] Muller op.cit fig.215;
[44] Muller op.cit fig.216;
[45] Muller op.cit p.250;
[46] Rankin op.cit fig.52;
[47] Wilkinson-Latham op.cit plate 82;
[48] Rankin op.cit figs.62 and 70;
[49] wysiwyg://107http://www.heroica.org/Shako.html;
[50] wysiwyg://107http://www.heroica.org/Shako.html;
[51] wysiwyg://107http://www.heroica.org/Shako.html;
[52] Delacroix: Liberty leading the People (lower right);
[53] Republique Francaise 24 February 1848;
[54] Rankin op.cit fig.44;
[55] Illustrated London News August 19, 1854;
[56] Wilkinson_Latham op.cit. fig.57;
[57] wysiwyg://107http://www.heroica.org/Shako.html;
[58] Illustrated London News August 11, 1860 p.331;
[59] Illustrated London News February 25, 1860 p.177;
[60] Waverly Lewis op.cit p.57;
[61] Blunderbuss Antiques item 13876;
[62] Blunderbuss Antiques item 13877;
[63] J.P. Morange;
[64] http://www.lehussard.fr/impression.asp?ar Ref=532066;
[65] http://www.lehussard.fr/impression.asp?ar Ref=533474;
[66] http://www.lehussard.fr/impression.asp?ar Ref=13008;
[67] http://www.lehussard.fr/impression.asp?ar Ref=129720;
[68] http://www.lehussard.fr/impression.asp?ar Ref=532020;
[69] Regimentals Lot 28638;
[70] Blunderbuss Antiques Item 11615;
[71] Muller op.cit pp.174-176;
[72] Tavard op.cit pp.224-225;
[73] L.Rousselot plates (cavalry helmets);
[74] Wilkinson-Latham op.cit p.63;
[75] Spink; The Goiris-Verhoef Collection of Militara Lots 324 and 325;
[76] Rowe David JJ: Head dress of the British Lancers 1816 – to the present fig.1;
[77] Carman op.cit p.70;
[78] Carman op.cit p.72;
[79] Rowe op.cit fig.18;
[80] Rowe op.cit p.50;
[81] Rowe op.cit Caps 7 & 8;
[82] Rowe op.cit p.97;
[83] Rowe op.cit fig.118;
[84] Rowe op.cit Cap 9;
[85] Muller op.cit fig.181;
[86] Johannson op.cit p.109;
[87] Johannson op.cit p.113;
[88] Spink op.cit figs.198 and 199;
[89] Waverly Lewis op.cit fig.67;
[90] Rankin op.cit fig.179;
[91] Johannson op.cit p.111-112;
[92] Knotel print published in Soares Branco op.cit p.61;
[93] Soares Branco op.cit p.63;
[94] Rowe op.cit fig 18;
[95] Soares Branco op.cit pp.69-75;
[96] Deighton Denis: The Battle of Waterloo;
[97] Rowe op.cit p.152;
[98] Rowe op.cit p.154;
[99] Blunderbuss Antiques item 17773;
[100] Regimentals item 19697 and item 19698;
[101] Regimentals item 19754;
[102] Regimental's item 25095;
[103] Spink op.cit Lots 197 and 198;
[104] Wallisand Wallis Spring Sale 1990;
[105] Rowe op.cit p.109-115;
[106] Muller op.cit p.78;
[107] Tavard op.cit p.152-160;
[108] The name busby could have derived from the name of wig worn by a headmaster of Westminister or the name of a british Colback manufacturer of the same name;
[109] Carman op.cit pp.54-68;
[110] Wilkinson-Latham op.cit p.54;
[111] Rankin op.cit p.93;
[112] Kube op.cit p.113;
[113] Johanssen op.cit pp.94-97;
[114] By L. Woolfe;
[115] The battle of Tyn was fought in 1744;
[116] The Illustrated London News Aug.19, 1854;
[117] Faughan Thomas: Stirring Incidents in the Life of a British Soldier p.410;
[118] Faughan op.cit p.372;
[119] The Illustrated London News Oct. 6, 1866;
[120] Johansson op.cit p.96;
[121] Regimentals Lot. 19474;
[122] Blunderbuss Antiques Lot 18109;
[123] Regimentals Lot. 26259;

7

The Pickelhaube

From Inkerman to the Somme

"But no! there, behind the Arc de Triomphe, there was a confused sound, a black line advancing in the growing daylight—then, little by little, the spikes of the helmets glisten, the little drums of Jena begin to beat, and under the Arc de l'Etoile, accompanied by the heavy tramp of the troops, by the clatter of sabers, bursts forth Schubert's Triumphal March."

Alphonse Daudet
The Siege of Berlin
1840-1897

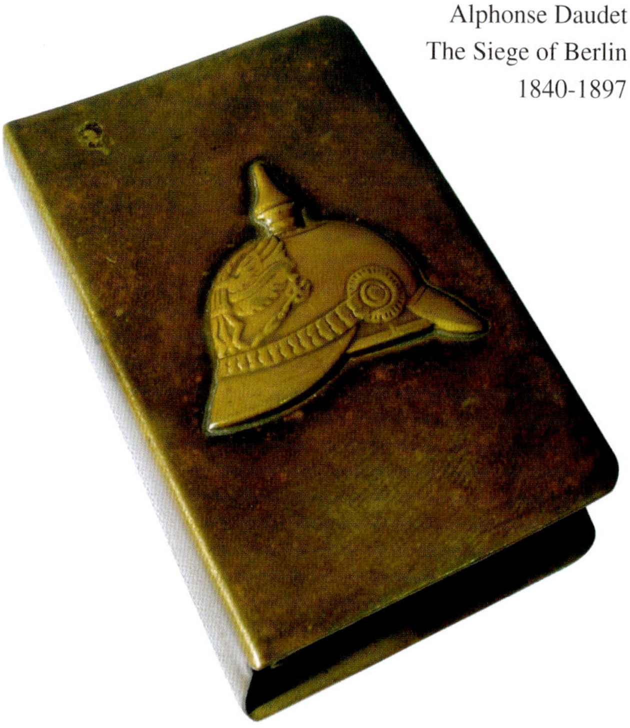

Patriotic German trench art matchbox case made from a spent cartridge c 1915. (Private collection - Malta) Photo by M Bonnici

October 23, 1842, witnessed the appearance of a helmet that would influence the evolution of military uniforms for almost a hundred years. 1842 witnessed the birth of the Picklehaube, an artefact that was destined to become the symbol of Imperial Germany. The Picklehaube was a sturdy boiled leather helmet surmounted by a tall brass spike and reinforced with brass fittings. It was a better helmet than the Shako. It was relatively stable on the head, and offered its wearer significant protection against sword cuts, but musket balls were another story.

The birth of a legend

Although several authoritative articles and books have been written on the Picklehaube, the origins of this distinctive type of headgear remain a mystery. There appear to be two schools of thought on the matter.

The first, 'traditional' school of thought endorsed by Rankin[1] sustains that King Friedrick Wilhelm IV copied the Picklehaube from a prototype helmet he noticed on the desk of the Russian Tsar Nicholas the First. According to this hypothesis the Russian military machine issued its Spiked Helmet only in 1844, while the more efficient Prussian military accomplished its task two years earlier in 1842.

The second school of thought confirms that the Picklehaube was commissioned by King Friedrick Wilhelm IV,[2] but according to this hypothesis it was the Russians who copied the Picklehaube from the Prussians, and not vice versa. Dates

Opposite: Artist's impression of a series of model 1842 Picklehaubes.

The Model 1842 Prussian Picklehaube

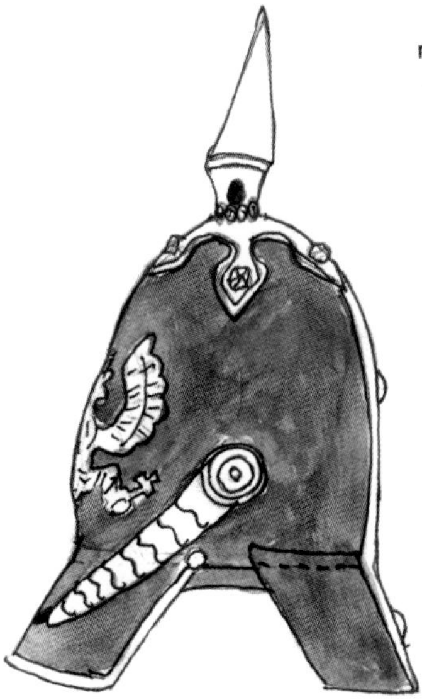

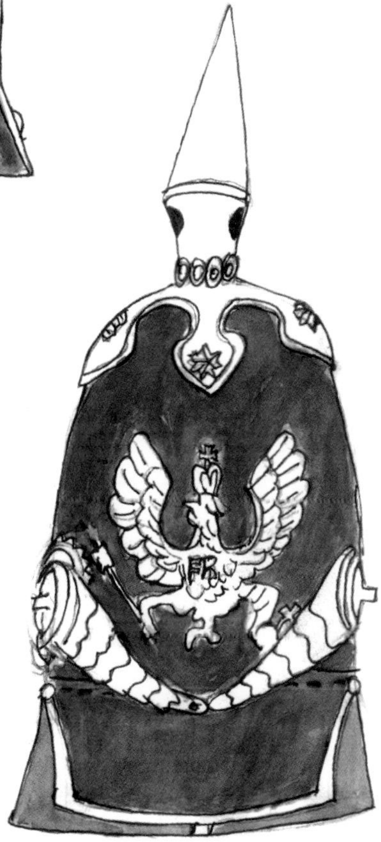

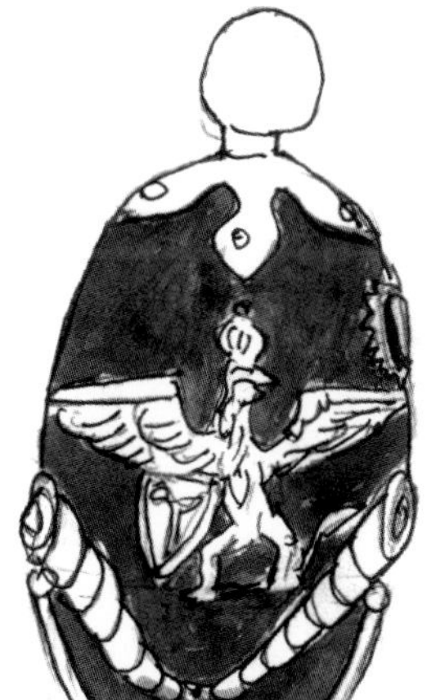

Baden Artillery

Swedish Infantry

seem to confirm this hypothesis, because the Russians adopted the Picklehaube two years after the Prussians.[3]

The author tends to agree with the 'traditional' school of thought for a number of reasons. The custom of surmounting helmets with spikes was a typical eastern tradition. Metal helmets fitted with spear point spikes were very popular throughout Asia Minor. In fact, the Persians still wore metal helmets fitted with spikes (known as Kulah Khudds) in the days that saw the birth of the Picklehaube. It is also very important to point out that Spiked Helmets were very popular in medieval Russia, and that the Spiked Helmet always appears on the head of the Russian national hero Alexander Nevsky in Russian represtentations. Whereas the Spike Helmet was alien to Friedrick Wilhelm's Prussia, the spike helmet was a characteristic of some Russian helmets, so it is possible that the spiked helmet was a Russian and not a Prussian contrivance.

7.2 The Picklehaube in Russia

It appears that the Russian Picklehaube replaced the Shako on May 9, 1844, when the infantry was equipped with an identical helmet to the one issued to the Grenadiers earlier on.[4] The Russian M1844 was a heavy and impressive leather helmet. It consisted of a tall leather dome reinforced by a pair of leather half moons that were stitched at the sides of the helmet. A pair of leather strips further strengthened the helmet. A leather brim (without brass bindings) was stitched to the front, while a

A 19th century etching illustrating members of the Imperial Russian Cavalry. (Private collection - Malta) Photo by M Bonnici

A 19th century etching illustrating members of the Imperial Russian Guard Curassiers. (Private collection - Malta) Photo by M Bonnici

The adoption of the Picklehaube created a problem in the Russian infantry. Officers and men were equipped with identical helmets, and the need was felt to introduce a special pattern helmet for officers. This problem was solved on January 4, 1845, when special fittings were approved for officers' helmets. It was decreed that national cockades (orange/white/orange) had to be fitted on officers' helmets under the chin scale boss. On January 9, 1848, officers and generals were allowed to

rectangular back peak protected the neck. A narrow comb supported the back of the helmet, while a large copper plate (with double headed eagle) reinforced the front part of the helmet. An imposing hollow spike supported by a cruciform base surmounted this massive headdress. The M1844 was around 36 centimetres high. It was fitted with a heavy-duty leather and beige cloth lining.

Father of all Picklehaubes? An M 1844 Other ranks Russian Picklehaube from the Crimean War. This interesting helmet was the subject of correspondence between the author and Eric Johannson. (Private collection - Malta) Photo by M Bonnici

This rare, early Russian Picklehaube of heavy construction appears to have been an arsenal re-issue. Photo by M Bonnici

The helmet plate of the M 1844 Russian Pickelhaube. It is a marine's helmet plate that was probably originally issued as a Shako plate. It shows significant signs of wear and damage. (Private collection - Malta)† Photo by M Bonnici

Side view of the Russian Picklehaube. Photo by M Bonnici

Inside view of the M 1844 Russian Picklehaube. Note the old style cloth and leather liner. (Private collection - Malta)† Photo by M Bonnici

Close up of the Russian helmet's profile. Note the reinforcing leather strips at the side. This feature was missing from the M1842 Prussian helmet. Photo by M Bonnici

substitute their hollow flame shaped 'spikes' with horsehair plumes on ceremonial duties (an exception was made in April 1848 for the officers of the King of Sardinia's Infantry regiment).

A lower and lighter helmet was introduced in 1860.[5] Unlike the earlier model, the crown of this helmet consisted of a single piece of leather. This Picklehaube pattern was substituted with an even lower helmet with brass trimmed edging in 1910.[6] It appears that this helmet was in use during the First World War, but the author failed to locate evidence of the use of the Russian Picklehaube on the eastern front.

7.3 The Picklehaube in Prussia

The Picklehaube was formally issued to the Prussian Army on October 23, 1842. This helmet was denominated the M1842. It was a very tall helmet (around 38 centimetres tall) that consisted of a tall leather dome with brass fittings. A square peak and a square brim were stitched to the dome. The brim was trimmed in brass. A leather cockade was worn on the right.[7] A spine strengthened the rear of the helmet and secured the back brim onto the dome. The M1842 was worn with convex brass chin scales.

A Bavarian Model 1848 Picklehaube of the Home Guard c. 1842-1873. *(Courtesy of Johan Somers)*

A lower helmet was introduced in 1856. This helmet was fitted with a leather cockade. A lower helmet was decreed four years later when a new range of helmet plates with the bandeau 'Mit Gott Fur Koening und Vaterland'[8] was decreed. Experience gained in the war against Austria in 1866 heralded some further modifications intended to ease mass production. The cruciform spike base was substituted with a round base, and the square brim was substituted with a round brim.[9] The leather spine was abolished, and the manner of securing the helmet plate to the helmet was changed.[10] Further changes were made subsequent to the Franco-Prussian war of 1870.

A rare M 1871/95 Saxon other ranks Picklehaube. This helmet is a model 1871 that was re-issued in the First World War as an M 1895 helmet. (Private collection - Malta)† Photo by M Bonnici

Side view of the M 1871/95 Picklehaube. Note the domed skull and the flared skirt of the M 1871 Picklehaube. (Private collection - Malta) Photo by M Bonnici

Detail of an ORs Saxon Picklehaube. The central device is made of German silver, while the starburst backing are made of copper. (Private collection - Malta) Photo by M Bonnici

Inner view of the Picklehaube. Note the multiple issue stamps on the skirt. (Private collection - Malta) Photo by M Bonnici

Rear view of the M 1871/95 Picklehaube. Note the M1871 style one piece back comb. (Private collection - Malta) Photo by M Bonnici

Detailed view of the M 1871/95 Picklehaube's Ebert maker's stamp and liner. (Private collection - Malta) Photo by M Bonnici

The rear spine was restored, and the bolt and nut manner of securing the helmet plate to the helmet was re-introduced. Curious and short-lived changes were made in 1887. All these innovations were made for reasons of economy, as the number of expensive brass components had to be reduced. The brass peak binding was done away with in favour of a strengthened leather peak binding in the old M1844 Imperial Russian style. Similarly, the brass chin scales were abolished in favour of leather straps for the troops. A new manner of securing the chinstrap to the helmet was also adopted.

Most of the 1887 measures were reversed in 1891. The leather chinstrap was retained, but a new manner of securing the strap system to the helmet (the M1891) was introduced. The brass brim binding was re-introduced, as helmets with M1887 peak bindings were found to be too fragile (the same deficiency had probably been experienced with the Russian M1844, because later models were issued with brass bound peaks). The M1891 changes also included a further reduction in the size of the helmet, giving the M1891 a slightly more squattish appearance. A ventilation hole in†the spine of the helmet was introduced in 1895.

A M 1915 Prussian Picklehaube with Garde eagle. (Private collection - Malta) Photo by M Bonnici

A standard M 1915 Prussian Picklehaube with pewter fittings. (Private collection - Malta) Photo by M Bonnici

Side view of M 1915 Prussian Picklehaube with pewter fittings. (Private collection - Malta)†Photo by M Bonnici

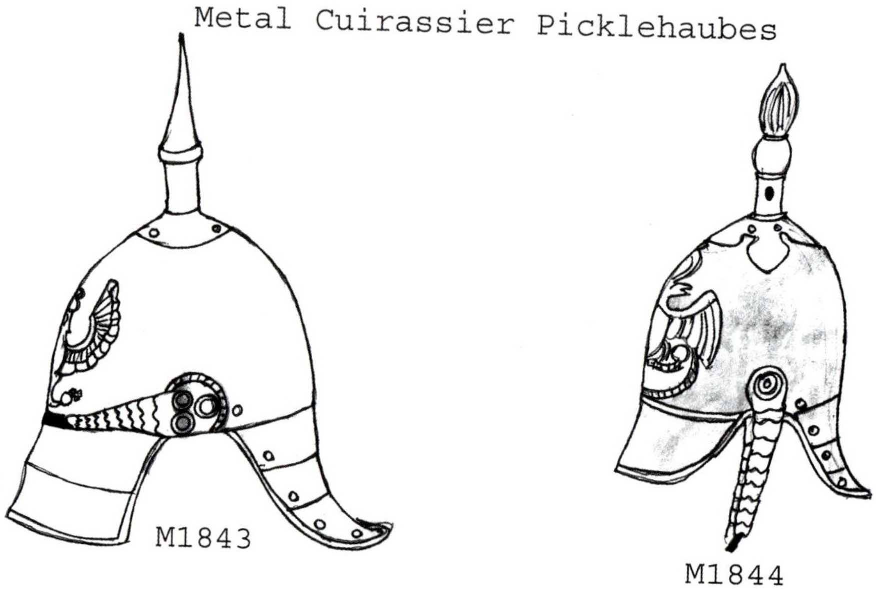

Artist's impression of M 1843 Model Cuirassier helmets.

Bavarian parade Picklehaube with horsehair bush c. 1910. *(Courtesy of Johan Somers)*

The evolution of the leather Picklehaube influenced the development of the metal helmet. Metal helmets of a (lower) profile similar to those of the M1856 were issued in 1867.

While metal helmets of a similar shape to that of the M1871 were issued in 1889. Cheaper patterns were issued in M1915 when Kurassier and Jaeger Zu Pferde helmets were issued with white metal fittings.[12]

A wartime issue (M 1915) Prussian Cuirassier's helmet (restored). (Private collection - Malta) Photo by M Bonnici

Side view of the M 1915 Cuirassier helmet. It has an unusually long lobster tail. Photo by M Bonnici

Inner view of the helmet. The brim retains its original light green finish. The liner is flaky but complete. Photo by M Bonnici

Close up view of the Armourer's mark inside the helmet. The mark of Lachmann is identifiable, but the date of issue is illegible. Photo by M Bonnici

An enlisted man's helmet of the Jaeger Zu Pferde (c. 1915). *(Courtesy of Johan Somers)*

The Picklehaube in Other Countries

The astounding victory of Prussia in the Franco-Prussian War inspired the admiration of many an army, and the uniforms and traditions of the warlike Prussians were copied in Europe and beyond.[13]

Britain adopted a Picklehaube style helmet (known as the 'Home service helmet') in 1879. The home service helmet was a blue cloth version of the tropical helmet used by British troops at the time. It was made of pressed cork covered with panels of blue cloth, and fitted with white metal and brass fittings. A slender spike that was more dangerous than that of the harmless Picklehaube rose from a crosspiece at the top of the helmet. The Light infantry was issued a grey cloth version of this helmet. Blue-cloth and grey-cloth helmets were retained in use up to 1920.[14] A similar helmet to the British blue cloth was adopted in the United States of America in 1881.

Ths helmet fell in disuse in 1902. The Omani Royal Bodyguard wore a similar helmet in scarlet cloth.[15]

Although Austrian infantry retained the Shako, it appears that a brass bound leather spike helmet of M1871 Prussian form with corrugated spike was issued to the Police and the *Guard du Corps* in the late 19th century.[16] Chile adopted a quasi identical spiked helmet to that worn by Hessen (with enamelled plate scrolled '*Con La Razon o La Forza*' in the same period). The Boer orange free state wore a facsimile of the Saxony Artillery Picklehaube.[17] The Mexican army adopted a Picklehaube of M1895 form fitted with a French Napoleonic style helmet plate. A cheaper version of this helmet is worn for parade wear today.[18] Some nations issued kitschy versions of the Picklehaube in coloured leather. The Republic of Estonia

Two U.S. blue cloth plumed Picklehaubes from the late 19th century. *(Courtesy of Tom Berndt, Militaria International Magazine).*

Above: A (badly) restored Spanish lancer's helmet from the Spanish American War. Some of the helmet's components were recovered from wrong U.S. M 1881 fittings.. (Private collection - Malta) Photo by M Bonnici Right: Close up of the Spanish lancer's plate. It appears to have been vandalised. Photo by M Bonnici

recently adopted an odd plumed version of the Prussian Picklehaube finished in glossy beige, while the Venezuela Presidential guard wears a Picklehaube lacquered in metallic blue.[19] Portugal and Spain wore much more elegant helmets.

Spanish lancers and cuirassiers wore a nickel facsimile of the Prussian M1889 cuirassier helmet. Very elegant Picklehaubes were used in Portugal. Portugal adopted the Picklehaube in 1860, but the Portuguese were not mere plagiarists, and improved the construction of the Picklehaube by introducing better quality fittings. The Portuguese Picklehaube also had brass bound peaks. A slightly shorter helmet was decreed in 1885. Ventilation holes were added on the sides of the helmet, while further austerity measures were made in 1902.[20] Similar Picklehaubes (made of fibreglass) are still worn today.

It is important to note that some states had adopted the Picklehaube before the outbreak of the Franco-Prussian war. The Italian state of Parma used a Prussian M1842 helmet in the 1850s, while Sweden used a facsimile of the M1842 in the 1850s (the evolution of the Swedish Picklehaube later closely followed the evolution of the Prussian Picklehaube, metal versions included).[21] Even the Vatican state adopted a tall black leather helmet similar to the M1842 in the mid-19th century.

A Portuguese Picklehaube c. 1930 in the Prussian manner. Photo by M Bonnici

Side view of the Portuguese Picklehaube. Most of the fittings are of a higher quality to the fittings mounted on German Picklehaubes. (Private collection - Malta) Photo by M Bonnici

Battles Fought in Picklehaubes

By some twist of fate the first battles fought in the Prussian M1842 were fought against Germans. Prussian troops wearing M1842 spike helmets crushed the revolution of 1848.[22] The Prussian M1856 was worn during the occupation of Shleswig-Holstein and the Austrio-Prussian war of 1866. The 1866 pattern was worn during the Franco-Prussian war, while all the later versions were employed on the western front in the First World War, the reason for this being the Germans' characteristic thrifty nature. It appears that when the Prussian military machine authorised a new pattern helmet the Prussians used to cut down and modify existing helmets to render them compatible with army regulations. The author had the opportunity to view an M1856 constructed from a cut down M1842. These modifications became very popular in the course of the First World War, when Germany was short of materials. The author's collection includes a rare M1871 dated 1873 that was re-issued as an M1895 by a replacement of the side-posts and spike.

Unlike the Prussian Picklehaube, most of the facsimile Picklehaubes worn by other armies around the world never saw active service. The British blue cloth helmet is case in point. However, the same may not be said of the Spanish metal cavalry spiked helmet which, judging from captured examples in U.S. collections, appears to have seen extensive use in the Spanish American war.

Evidence of the use of the Russian M1844 abound.[23] Excerpts from the illustrated London news suggest that this artefact was a very popular war trophy[24] with British Veterans.

A Crimean war period etching that illustrates a Russian infantryman wearing what is meant to be an M 1844 pattern Picklehaube.). (Private collection - Malta) Photo by M Bonnici

An M 1895 Prussian other ranks Picklehaube with brass fitting..
(Private collection - Malta) Photo by M Bonnici

Above: A bullet hole cuts right through the Picklehaube's helmet plate. Below: View of the M 1895 Picklehaube liner. Photos by M Bonnici

Rear view of the helmet. Note the bullet's exit hole. Photo by M Bonnici

A relic M 1895 Bavarian officer's Picklehaube from the Western front. (Private collection - Malta) Photo by M Bonnici

A Model 1897 Prussian Line Cuirassiers Parade helmet. *(Courtesy of Hermann Historica)*

An enlisted man's helmet of the Empress Maria Feodorovna Horse Guards C. 1900. *(Courtesy of Hermann Historica)*

A helmet and cuirass of the Maria Feodorovna Horse Guard regiment c. 1900. *(Courtesy of Hermann Historica)*

A Russian model 1852 helmet of the lifeguard Horse Guard Regiment with a modified late French cuirassier associated cuiraiss. *(Courtesy of Hermann Historica)*

8

The Bicorn Hat

From battle hat to dress hat

"At that moment Stepan Arkadyevitch, his hat cocked on one side, with beaming face and eyes, strode into the garden like a conquering hero. But as he approached his mother-in- law, he responded in a mournful and crestfallen tone to her inquiries about Dolly's health. After a little subdued and dejected conversation with his mother-in-law, he threw out his chest again, and put his arm in Levin's."

Leo Tolstoy
Anna Karenin
1877

The Bicorn Hat (also known as a Chapeau) evolved from a late variant of the Tricorn Hat. The Bicorn hat was very similar to the Tricorn hat, save for the fact that it had two horns instead of three. The earliest examples of this hat pattern date back to the fourth quarter of the 18th century,[1] when a hat with only a pair of flaps cocked inwards appeared in Prussia. The shape of this hat was very similar to that of the Tricorn, but on the Bicorn the flaps that were cocked inwards were the front flap and the back flap. The edges of these hats were reinforced with a band of cloth tape. Versions worn by officers were reinforced with gold and silver braiding, and edges were trimmed with heron

A 19th century lithograph illustrating Napoleon I, Napoleon II, and Napoleon III wearing Cocked Hats. (Private collection - Malta)† Photo by M Bonnici

plumes.[2] Towards the end of the 18th century Cocked hats with higher back flaps became very popular, and the Cocked Hat developed into the *'Petit Chapeau,'* associated with Napoleon Bonaparte. Around thirty years after the fall of Napoleon the Bicorn Hat began to be conceived purely as a parade hat, and a special type of Chapeaux known as the Chapeaux du bras came into being.

8.1 The Chapeau

"*Wherever the party stopped, and an opportunity was offered, Mr. Jos left his own card and the Major's upon "Our Minister." It was with great difficulty that he could be restrained from putting on his Cocked Hat and tights to wait upon the English consul at the Free City of Judenstadt, when that hospitable functionary asked our travellers to dinner. He kept a journal of his voyage, and noted elaborately the defects or excellences of the various inns at which he put up, and of the wines and dishes of which he partook....*"

William Makepeace Thackeray
Vanity Fair
1848

The Cocked Hat that was destined to become synonymous with the most famous Frenchmen of all was officially issued to the French military force in 1793.[3] In France the Cocked Hat, which was known as a Chapeau, was worn cross wise over the head. The M1793 Bicorn was a soft felt hat that was rather shapeless. This serious deficiency signalled the death knell of the Bicorn hat, and an 1806 order substituted the Bicorn hat with the Shako. However, the Bicorn was retained by chosen corps and by certain members of the officer class. A special Bicorn was issued to the cavalry in 1801. The front flap of the M1801 was slightly shorter (17.6 centimetres) than the back flap (19.6 centimetres). Following the introduction of the neo-classical helmet, some members of the Imperial French cavalry retained the M1801 for undress wear. A glorious version of the M1801 with bands of gold lace was issued to the general officers in 1803. This hat was the hat worn by most of Napoleon's marshals.

A special Cocked Hat known as *'petit Chapeau a' la Francaise''*was worn by Napoleon Bonaparte. Napoleon's austere Cocked Hat was much less ostentatiously adorned than the hat worn by his marshals. It was a taller version of the

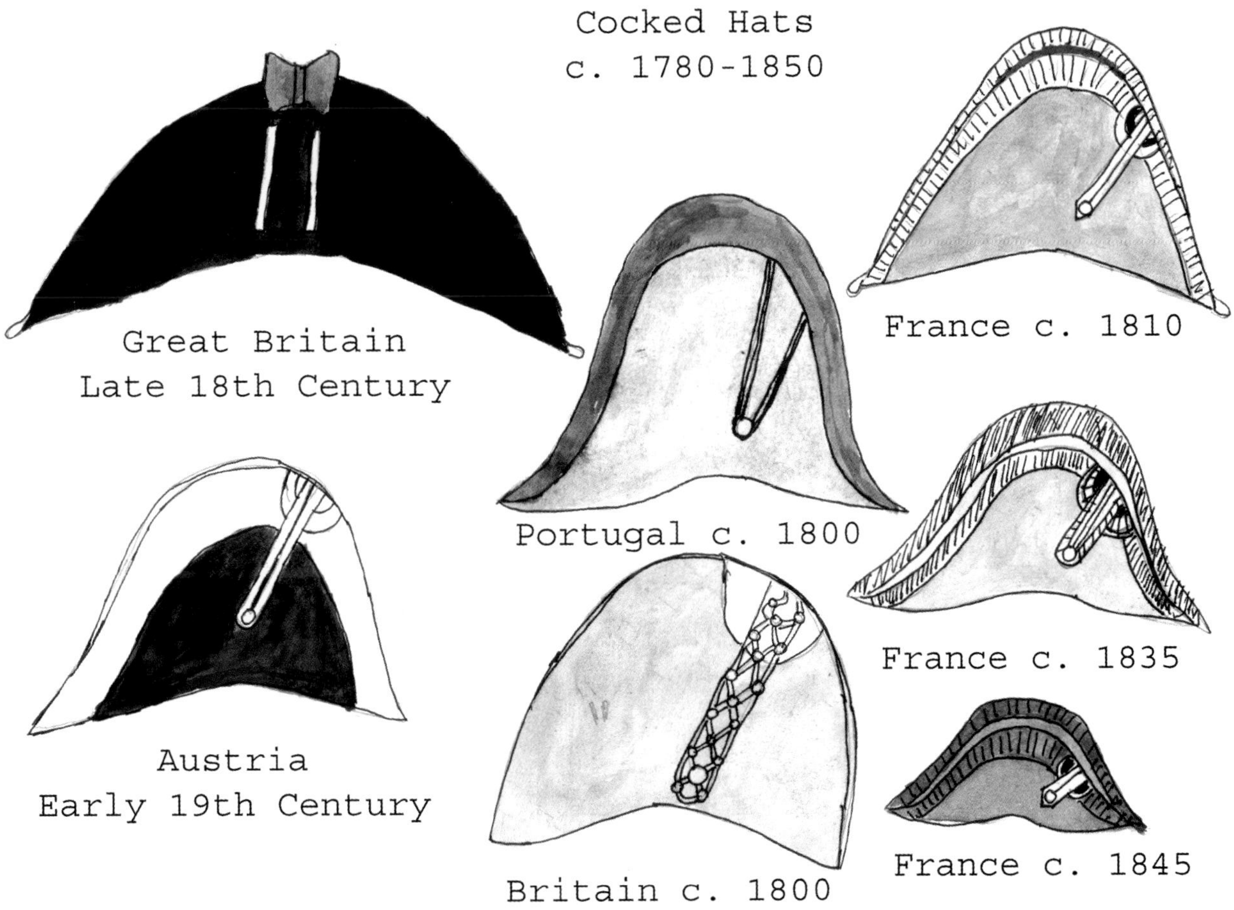

Artist's impression of a number of 19th century Cocked Hats.

A rare French (officer's?) Cocked Hat c. 1835 (possibly Gendarmarie). (Private collection - Malta) Photo by M Bonnici

Side view of the French Cocked Hat. Photo M Bonnici

Inside view of the French Cocked Hat—the liner is torn. Photo M Bonnici

Rear view of the French Cocked Hat. Photo M Bonnici

M1801, as worn by officers for undress wear. It featured no expensive gold braiding; and decoration was restricted to a cockade with the national colours. It appears that Napoleon made use of around 120 hats of this pattern during his years of fame, and that he used two hatters, Poupart and Maneglier. A relatively significant number of hats that were worn by Napoleon survived to the present day. The hat worn by Napoleon at Waterloo is kept at the Zeughaus in Berlin, while the hat worn by Napoleon at Saint Helena is kept at the *Musee' de la Ferme du Caillou*. Approximately†five hats that belonged to Napoleon are kept at the *Musee' de' l Armee*.

The M1803 Chapeau was abolished with the fall of Napoleon. A new style of Chapeau was introduced in 1815. It was very similar to the tall and bombastic Chapeau worn by officers in the last days of the empire, but unlike the First Empire Chapeaux, the M1815 Chapeau had an arched crown that mirrored the shape of the Chapeau's back flap. This Chapeau style was worn up to M1835, when it was substituted by another Chapeau that was approximatley 25 centimetres at the highest point.[4]

The size of the Chapeau was reduced once more in 1844. The M1844 was a felt miniature Chapeau that was built on a cardboard framework. Unlike other Chapeaux, it was worn *en colonne* (with the points fore and aft), and the rigid upturned horns were kept in place by an iron framework.

It was significantly shorter than the M1803 and M1815 Chapeaux. Whereas M1803 and M1815 Chapeaux could rise up to almost 30 centimetres in height, M1844 models rose up

An unusual M 1846 French Cocked Hat. This hat appears to have been issued at the time of the second empire and re-issued after the fall of Napoleon III. (Private collection - Malta) Photo by M Bonnici

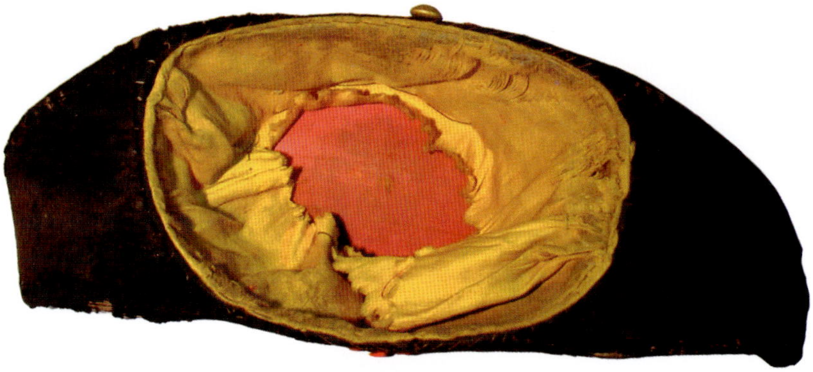

Inside and rear view of the Cocked Hat. Note the red stitching. Photo M Bonnici

to only around 20 centimetres.[5] An even shorter Chapeau (11 centimetres at the highest point) was introduced in 1872.[6] This hat pattern was used for parade purposes at the French military academy up to 1958. A similar hat was issued to the French navy and was used as a *Chapeau du Bras* up to 1930.[7]

The origins of the British Cocked Hat trace back to an order passed in July 1787, when a version of the Ramillies cock (the Tricorn Hat) with only two flaps was issued to all the branches of the British military. In the beginning of the 18th century taller Cocked Hats that assumed an imposing fan shape became very popular. The Cocked Hat was no longer worn crosswise but with the points fore and aft the head. This hat was very popular in the Royal Navy, and was worn by the heavy cavalry up to 1812. At this point in time it was still conceived as a functional hat, and was strengthened by an intricate net of braiding that was meant to offer protection against sword cuts.[8]

Similar hats were worn in Sweden and the German states,[9] but after the Napoleonic wars the Cocked Hat developed into the Chapeau du Bras and was no longer worn on the battlefield.

A British Victorian naval Cocked Hat. (Private collection - Malta) Photo by M Bonnici

A photo of the hat in its original storage tin. Photo M Bonnici

Inside view of Cocked Hat. (Private collection - Malta)†Photo by M Bonnici

The box of the Victorian Cocked Hat. (Private collection - Malta)Photo by M Bonnici

8.2 A Chapeau for the hand

"This was done so that all the army might be under arms to salute their chief as he passed. On these occasions he wore his dress uniform, Cocked Hat, aiguillettes, sabre and spurs. His staff proper, besides all officers constructively on his staff—engineers, inspectors, quartermasters, etc., that could be spared—followed, also in uniform and in prescribed order. Orders were prepared with great care and evidently with the view that they should be a history of what followed."

Ulysses S. Grant
Personal Memoirs
1885

All the military hats mentioned in this book broadly served one purpose—that of a hat covering—but the history of military headgear incorporates a hat that was not primarily designed as a headdress. The popularity of the bombastic French M1815 rapidly spread around the western world, and the tall hat became an indispensable component of gala dress. Officers were expected to present themselves with their tall Chapeau compressed under their arms. These hats became known as 'Chapeaux du Bras' (hand held hats),[10] and as time went by the designs of these hats became so flamboyant that these hats could no longer be worn on the head. By the mid-19th century almost all the military, naval, and administrative officers of the Western world were equipped with French M1844 pattern Chapeaux du Bras. Few military powers used these hats for active services, and these fine, small hats became a part of formal or ceremonial dress. Similar hats were worn in Italy, Bavaria, Prussia, Great Britain, Italy, France, Belgium, the Netherlands, Imperial Russia, Spain, Portugal, Sweden, the United States, and many other countries. The Cocked Hat patterns worn in these

Above: A rare Russian Cocked Hat c. 1900. Left: A close up of the Czarist cockade. (Private collection - Malta) Photos by M Bonnici

Inside view of the Russian Cocked Hat. Note the maker's mark on the silk lining. This hat was made in Helsinki. Photo M Bonnici

Close up of the Czarist orange cockade and the gilt button's twin headed eagle. Photo M Bonnici

A French Naval officer's Cocked Hat c. 1870. (Private collection - Malta) Photo by M Bonnici

Inside view of the Cocked Hat. Note the Toulon hatter's mark. Photo M Bonnici

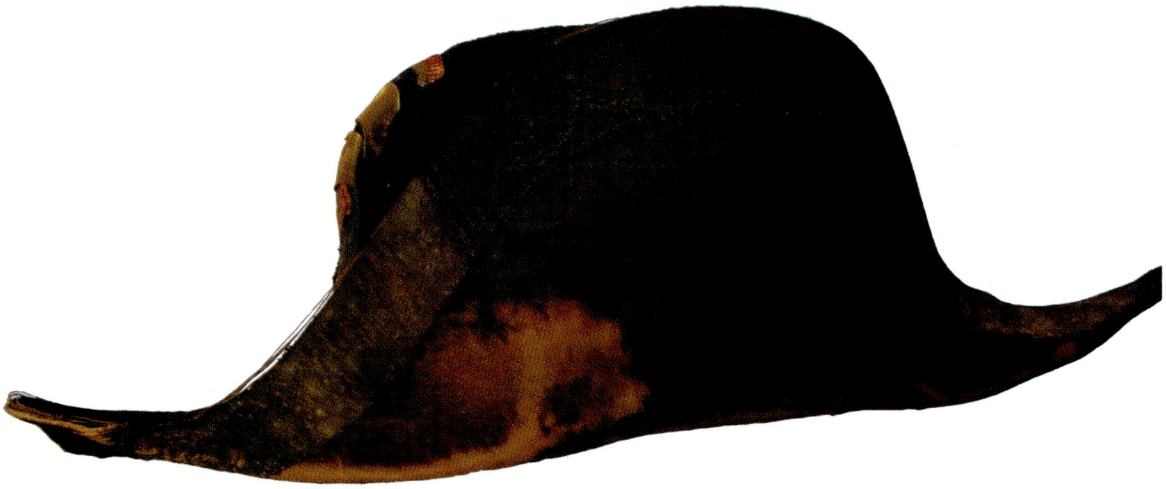

Rear view of the Cocked Hat. Note the signs of wear in the area where the hat was held. Photo M Bonnici

countries were quasi-identical; the only mark of distinction was frequently the button and the colours of the cockade. These hats fell into disuse with the outbreak of the Second World War, but countries like the United Kingdom still use these hats on ceremonial occasions.

Battles fought in Cocked hats

The history of the Cocked Hat is intimately knotted to the Napoleonic age. As Wellington's statue at Hyde Park reminisces, Lord Wellesley wore a Cocked hat as he galloped through the lines of his troops at Waterloo. Similarly, Lord Nelson's monument at Trafalgar Square depicts the hero of Aboukir bay wearing a hat of the Chapeau family. Needless to say, the Chapeau was Napoleon Bonaparte's hallmark. At the time the Cocked hat saw universal use amongst officers and dignitaries., so much so that Krafft's painting of the battle of Leipzig depicts Napoleon's foes—Frances I of Austria, Alexander the I of Russia, and Frederick Wilhelm III—wearing Cocked hats. Napoleon III later wore a similar hat to that worn by Napoleon Bonaparte, and William IV wore a similar hat to that worn by Lord Nelson.[11] Period illustrations of the French officers who fought the Crimean War depict them wearing the M1844 Bicorn,[12] a facsimile of the Petit Chapeau worn by Napoleon. This hat was still worn at the time of the intervention in Italy and the Franco-Prussian War.[13] The officers of the British Royal Navy wore these hats throughout the nineteenth century during active service, and used it for parade wear in the 20th century.

Collecting

"He made many attempts to induce me to consent to an exchange; at one time coming out with a fishing-rod, at another with a fiddle, at another with a Cocked Hat, at another with a flute. But I resisted all these overtures, and sat there in desperation; each time asking him, with tears in my eyes, for my money or my jacket. At last he began to pay me in halfpence at a time; and was full two hours getting by easy stages to a shilling."

Charles Dickens
David Copperfield
1850

The Cocked Hat is probably the least popular hat described in this book. Broadly, the run of the mill Cocked Hat is common.

A 19th century lithograph illustrating Lord Raglan wearing a Cocked Hat in the Crimean War. (Private collection - Malta) Photo by M Bonnici

The reason for this is that, unlike other headgear, most Cocked Hats were never subjected to the tribulations of war and were used exclusively as parade hats, resting hidden in their carrying cases for most of their lifespan. Consequently, the prices fetched by most post-1850 Cocked Hats are low when compared to the prices fetched by Neo-Classical helmets and Picklehaubes. Most of the Cocked Hats that appear on the collector's market are Naval Cocked Hats from the first quarter of the 20th century. The manner of construction and general appearance of these hats is very similar to that of their 19th century counterparts. Dating features on Cocked Hats include the height (the shorter the hat, the later the date of production), button (Victorian Cocked Hats have buttons wih Queen's crowns, while later pre-Elizabethan Cocked Hats have King's crowns), and the tailor's label on the lining (most 19th century hatters went out of production in the 20th century). A significant number of Cocked Hats, complete with their original carrying cases, have survived and regularly appear on the collectors' market.

As discussed above, it is only the later specimens that sell for relatively cheap prices. Hard to find early Cocked Hats fetch significant prices. Napoleonic war period and early Victorian Cocked Hats are rather expensive. In July 2003 a British dealer offered an extremely rare naval Bicorn dating circa 1800 at USD 5,120.[14] Other early Bicornes fetched high prices at internet auctions. The author recalls noticing a splendid Napoleonic war period officer's Bicorn hat being sold at USD 5,000, and an early Victorian naval hat being sold at USD 585. Later Bicornes are normally cheaper. In July 2003, an Australian dealer offered an Italian naval fore and aft hat being sold at USD 375.[15] In 1987, a British auctioneer[16] sold a fine Imperial Russian officer's Cocked Hat at around USD 370. In September 2003 a French dealer offered a Bicorn hat of a military attendant at around USD 200, a third empire Bicorn of a naval officer at around USD 370, and a third empire Bicorn hat from the military school at around USD 330.[17]

Replicas of early Cocked Hats are in circulation. The author noticed some highly suspicious Napoleonic first empire Bicornes on the market. The latter had slightly irregular patterns, were made from light felts, and did not have genuine aging characteristics. An important aging characteristic on well-aged, unboxed Bicornes is the colour of the felt; the back felt tends to discolour to a light grey with time. Collectors of early Cocked Hats are also faced with a more dangerous menace. Early Cocked Hats from the time of Louise Philippe are frequently stripped of their original buttons and fitted with First empire fittings. The appearance of these hats would deceive the most discerning collector, and only a thorough knowledge of the evolution of French Cocked Hat designs can arm a collector with the requisite knowledge to identify these† 'forgeries.' It is also important to note that the military was not the only cadre to make use of Cocked Hats in the early 1800s. In the 19th century undertakers, civil servants, and court officials wore Cocked Hats. The hat patterns worn by the latter were frequently quasi-identical to those used by the military, and the fittings of Cocked Hats should be thoroughly examined so as to ensure that the fittings on the hats are not recent additions. Replica first empire Bicornes were in circulation as early as 1840, and an aspiring collector of early Cocked Hats should be extra careful prior to diving into an expensive purchase. Later[18] Cocked Hats present much lesser concerns. At times, original late Cocked Hats are so inexpensive that it is probably not worth producing replicas of these good quality hats. However, there appears to be a tendency to confuse fraternal society and other non-military hats with late Cocked Hats, and collectors should be aware that no military hat was ever emblazoned with the scarlet cross of the knights of Colombus, nor fitted with the cut steel chains characteristic of British Lord Lieutenant Bicornes.

Notes:

[1] Kube op.cit fig. 28;
[2] Muller op.cit. fig. 227;
[3] Prior to 1793 this hat was used by some militia units and was known as *Bicorne des volontaires.*
[4] Margerand Joseph: Les Coiffures De L'Armee Francaise p.411;
[5] Margerand op.cit p.419;
[6] Margerand op.cit p.422;
[7] Tavard op.cit pp.123-138;
[8] Carman op.cit p.8-10;
[9] Wilkinson-Latham op.cit p.43;
[10] Rankin op.cit p.10;
[11] Martin Archer Shee: William the IV;
[12] The Illustrated London News August 19, 1854;
[13] Tavard op.cit pp.123-138;
[14] Blunderbuss Antiques Item 13870;
[15] Armor Antiques http://www.web.smart.net.au/gacton/armorantiques/ArmorPage/HE-Item12.htm;
[16] Wallis & Wallis Sale 318 lot 397;
[17] Le Hussard Bicorne Lots 1-4;
[18] Post circa 1840;

9

The Pith Helmet

From the Indian Mutiny to Gallipoli

"*For clothing Kermit and I used what was left over from our African trip. Sun helmets are best in the open; slouch-hats are infinitely preferable in the woods.... The traveller who enters tropical South America should protect his head with a wide-brimmed soft felt hat with ventilated headband, or the best and lightest Pith Helmet that can be secured, one large enough to shade the face and back of neck. There should be a ventilating space all around the head-band; the wider the space the better. These helmets can be secured in Rio and Buenos Aires. Headnets with face plates of horsehair are the best protection against small insect pests. They are generally made too small and the purchaser should be careful to get one large enough to go over his helmet and come down to the breast.*"

Theodore Roosevelt
Through the Brazilian Wilderness
1914

In the closing months of 1858, the British hatter Hawkes & Co.[1] of 14, Piccadilly, London, announced its latest patent, a revolutionary helmet made of lightweight cork. The new helmet was christened the Pith helmet.

Hawkes' announcement in *The Malta Observer*[2] publicised a new type of hat created '*to supply the great want of a proper covering for the head in tropical climates*.' The helmet claimed to be sun-proof and water proof.[3] The novelty of the brainchild lay in the fact that, unlike the other helmets available at the time, it was made of lightweight cork. The advert read '*this desideratum is fully obtained by Helmets, Hats and Caps, being made in any shape of Cork, under a new process by which it is rendered flexible, and its natural brittleness effectually removed. Every kind of head-dress covered with any material desired can be made exceedingly light or of great strength, sufficient to withstand any blow, with little extra increase of weight.*'

The ad reported that Cork Shakos were in the course of being manufactured by the British Government for a portion of the British Army in England. Hawkes' Pith Helmets were being sold for 25 shillings.

Hawkes & Co. claimed to be the sole manufacturers of this type of Helmet, and it warned:

"*A spurious imitation of this manufacture must be guarded against, as it is entirely destitute of all the advantages possessed by Hawkes & Co.'s Helmets, &c, which have their name stamped upon them.*"

The sole Agents for the Hawkes Helmets were T. R. Lewis of 16, Gold Square, London.[4]

Black Cloth Covered Cork Helmet, with polished brass chain strap; eagle spike and base; assorted sizes. Good serviceable order. Original cost upwards of $5.00 each. Our bargain price, $1.10.

500 Black cloth-covered Cork Helmets, with horse-hair plumes, polished gilt brass spread eagle, chin strap and plume holder; all in good second-hand cleaned order; assorted sizes; very showy helmet. Price, $1.45 each.

600 white Cloth-covered Cork Helmets, as used by State Militia, with polished brass gilt spike, spread eagle and chin strap; all in good cleaned second-hand order. Price, $1.10 each.

5000 New York State Militia Black Cloth-covered Cork Helmets, with gilt brass spike and base and gilt front ornaments; assorted sizes; all in good, cleaned, second-hand order, like new. Price, 85 cents each.

8,000 Militia Helmets, cork body, covered with fine black colth; almost new goods, assorted sizes; gilt burnished ornaments, spike, chain and eagle; handsome, showy helmets; $1.10 each; same in white cloth covered; with horse hair plumes, any color, $1.45 each.

U. S. Army Tan-colored Cotton Cloth Covered Cork Helmets; late regulation; cost the U. S. Government $1.24 each; assorted sizes. Our price for single helmets, 62c; by the dozen lot, $7.00.

British Army Helmet, fine black cloth, finished with gilt chain strap with spike and base. With any selection of British Army helmet plates advertised in our catalogue. Price, $3.45.

German Soldier's Chinese Campaign Helmet, with fine gilt bronze Prussian eagle ornament. Back of helmet folds up, as desired. Genuine relics, purchased by our agent in Germany on the return of the troops from China. Useful, ornamental relic. Price, 85 cents.

British Police Helmet, with brass front plate with letters and numbers, second-hand; serviceable. Price $2.75.

300 New U. S. Army Light Weight Helmets for use in tropical climates, made of fine light fibre, covered with light silver-gray cloth, with ventilated sweat band, good value for $4.00, all new, our bargain price, 90 cents each.

Puggaree Helmets either White or Khaki covered, regular U. S. A. Cork Helmet, with band of same colored cloth around the base; for stage use. New Helmets, assorted sizes. Price, $1.00 each.

6,000 U. S. Army New White Helmet, made of selected long light cork covered with fine white drill, with ventilator in top that can be unscrewed, and gilt ornamented spike, used if desired, with sweat band ventilator; assorted sizes. Price, 75 cents each, special price for export quantities.

Theatrical or Political Campaign Hats, with colored cord; no ornaments; 30 cents each, $3.00 dozen. New goods, assorted sizes.
Spanish soldier's Dress Hats; fine new goods; relics. $4.75 each.
Helmet Plumes, U. S. Army, late regulation; yellow, 40; white or red, 60c.
U. S. Infantry Officer's Helmet, with white yak plume, gilt ornaments and cord; size 7; new. Price, $10.00. One slightly used, price $7.50.

New U. S. Army Straw Helmet, made for use in tropical climates; rare, serviceable, from gov't auction. Price, $1.00 each.
Tourists or Campaign Hat, India style, in white and in drab colored; new goods, assorted sizes; 90 cents each.

German Army Leather Helmet, with brass ball, ornamented top; with German eagle with gilt chain scale strap. Good second-hand serviceable order. Price, $5.00.

British Police Helmet, stiff body cloth covered with metal ornament, denoting station. Illustration is (Dock Dept.) genuine "Bobby's 'At." Second hand, assorted sizes. Price, $2.85 each.

Prussian Body Guard Soldier's Leather Helmets, with brass spread eagle plate, spike, chain, etc., all in fine order, like new. Price, $3.75; 100 in stock.

Church—The old general always wanted to be where the fighting was thickest.
Kotham—Is that a fact?
Church—Oh, yes; why, even when he went to church he asked if he might sit in the choir.
—*Yonkers Statesman.*

A clipping from a 1927 Bannerman's (Military Surplus) Catalogue. It attests that spike helmets, Blue cloth helmets, and Pith helmets have been collected for a long time. (Private collection - Malta)† Photo by M Bonnici

A rare early Pith helmet c 1860. It was fitted later with Blue cloth helmet type mounts. (Private collection - Malta)†
Photo by M Bonnici

This helmet was subjected to heavy wear. It appears to have been painted or tarred at some point in time. It could be a stripped foul weather Pith helmet. Photo M Bonnici

Side view of this worn early British Pith helmet. Photo M Bonnici

View of the Victorian Pith helmet's liner. Photo M Bonnici

9.1 A Helmet for the Colonies

Hawkes' innovative helmet was destined to become an epitome of colonial Britain. It served in India, Malta,[5] Asia Minor, and North Africa.[6]

The Pith helmet was a superb helmet for its time. It was an excellent source of protection against the remorseless equatorial sun,[7] a terrible enemy to the fair members of the British army. Yet it suffered all the deficiencies suffered by most of the infantry helmets of the period.[8] Despite its sturdy appearance, the Pith helmet supplied its wearer with no serious protection. The most it could do was to glance off a Zulu Knobkerrie or the angry fist of a protester during a riot.[9] Otherwise, Dervish bullets, Sudanese swords, and African spears could pass right through the so-called 'helmets.'

History developed various types of Pith helmets,[10] and the appearance of the hat changed as time went by. In the mid-Victorian era, the Pith helmet was rather tall and dome shaped; it had a high crown and short peaks. This helmet pattern must have later influenced the development of the Blue Cloth helmet, which had a quasi-identical shape. The most notable difference from the Blue Cloth helmet was the colour; overseas helmets[11] were white or khaki, whereas Home Pattern[12] helmets were dark blue.

Two types of Pith helmets were worn in the Victorian era: the Officers' version and the Troopers' version. The Officers' version was embellished with shining brass fittings. It often featured a spike and a floral spike base. A sturdy brass trim strengthened the edges of the Officers' helmet, while the roped chin chain and occasional surmounted Heron plumes gave the Officer helmets a touch of elegance.[13] Officers' helmets were generally privately purchased, unlike the examples worn by troopers, which were issued, free of charge, by the Crown.[14]

The helmets worn by troopers were much less ornate and fine. They were mass-produced under Government contract. Chinchains were replaced with leather straps, and most examples featured no fittings at all. The interior of Troopers' helmets was also of a lower quality to that of the expensive Officers' helmets.

The rough tan Pith helmet reached the zenith of its popularity in the closing years of the 19th century and the opening decades of the 20th century.

An illustration from the Spanish American war period. It depicts U.S. troops wearing M. 1872 Kepis and Pith helmets of British inspiration. (Private collection - Malta) Photo by M Bonnici

The Dardanelles was plagued with a torrid desert climate, and the authorities decided to equip the Gallipoli army with Pith helmets.[15] By the time of the Gallipoli operation, the Pith helmet had evolved into a different shape from that of its Victorian ancestor. The Pith helmet of the First World War had a large brim and an ungainly protruding neck flange, and was known as the Wolesely pattern. It looked flatter and wider than its Victorian predecessor. The Belgian army made use of a similar helmet in the Congo.[16]

All nations that had colonial aspirations donned an Army equipped with Pith helmets. Despite the poor show at Gallipolli, Pith helmets continued to be worn long after the end of the hostilities of the Great War. They were worn throughout the Second World War by Allied and Axis powers alike.[17]

In the inter-war years a smaller version of the Solar Pith helmet (the Bombay bowler) picked up popularity. The origins of this type of helmet may be traced back to the years of the late Victorian African Safaris.[18] Women used special versions of this squattish helmet.[19]

A British First World War 'Wolseley Pattern' Pith helmet in unissued condition (without pugree). (Private collection - Malta) Photo by M Bonnici

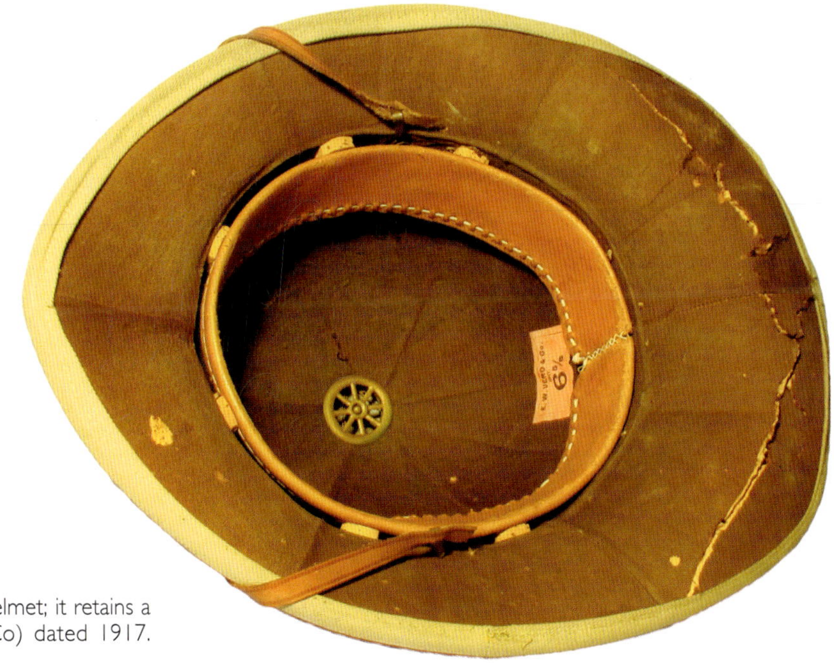

Inner view of the Pith helmet; it retains a tailor's label (Vero & Co) dated 1917. Photo M Bonnici

Battles Fought in Pith Helmets

Most of Britain's colonial wars were fought in Pith helmets. The Pith Helmet saw over one hundred years of service, and was employed in some of the most important battles in colonial history. The Indian Mutiny was one of the earlier campaigns that witnessed the use of the Pith Helmet on the battlefield.[20] The British Pith Helmet was later used in the Abyssinian campaign of 1867-68,[21] the Ashanti War,[22] the Afghan War of 1878-81,[23] the Egyptian War of 1882,[24] the North West frontier,[25] the Boer War,[26] and the Gallipoli Campaign.[27] It was after used in numerous Second World War battlefields. The popularity of all forms of Pith Helmet suffered a decline in the post-war, but the Pith Helmets did not run completely into extinction.[28]

It is important to note that the Pith Helmet was not the only form of colonial headgear worn in the late 19th century and the first half of the 20th century. Colonial empires developed a plethora of distinctive headgear styles. The *Turcos*[29] of the French Army wore the scarlet *Chechia*, while Britain's Indian troops wore a Maharajah style turban. The slouch hat (known as the *Hut* in Prussia) became very popular in the Unites States and Prussia after the Spanish-American war, while other forms of characteristically ethnographical hats were worn by the auxiliary troops of the major colonial powers.

9.3 Collecting

The Pith Helmet appears not to be a very popular helmet among collectors. It enjoys some popularity in countries with a colonial heritage, such as Britain, but the popularity of this functional helmet is not comparable to that of the Picklehaube or the Neo-Classical helmet. However, most 19th century Pith Helmets are rather rare, and the rarity of these helmets has a bearing on their prices. In September 2003 a British dealer offered a 1878 pattern Officer's Pith helmet complete with original khaki combat cover and straps at USD 2,310,[30] a Boer war period officer's foreign service helmet at USD 1,815,[31] and a well documented Boer War period Glamorgan Yeomanry tropical helmet formerly owned by Maj.Venables Llewelyn at USD 2,026.[32] Later 'Wolesely' pattern helmets are much cheaper. The same British dealer offered a documented First World War period Pith Helmet at USD 536, a World War One Pith Helmet at USD 486, and a 1920s Pith Helmet at USD 280.[33]

Enormous quantities of replica Pith Helmets are being produced for the re-enactor market. Some of these are made of canvas covered fibreglass, are very light, and do not represent a menace to the average collector. The same may not be said of replicas made from pressed cork using traditional methods of Pith Helmet production. The latter may be a concern to collectors, but specimens noticed by the author do not have the crisp characteristics of the originals: the contours are too sharp; the seams are of a poor quality; and the shape is†occasionally irregular.

A 1915 photograph published in War Illustrated. It depicts Allied forces wearing Wolseley pattern Pith helmets. (Private collection - Malta)†Photo by M Bonnici

EQUESTRIAN PORTRAIT OF OMAR PASHA, COMMANDER OF THE TURKISH FORCES.

A 19th century lithograph that illustrates Ottoman Cavalrymen wearing a fez type hat similar to the Chechia worn by the *Chasseurs d'Afrique*. (Private collection - Malta) Photo by M Bonnici

Artist's impression of tropical headgear worn by the colonial forces of various empires, and of a series of colonial headgear.

Headgear worn in the Colonies or worn by the Colonial Troops of the Major Empires c. 1870-1918

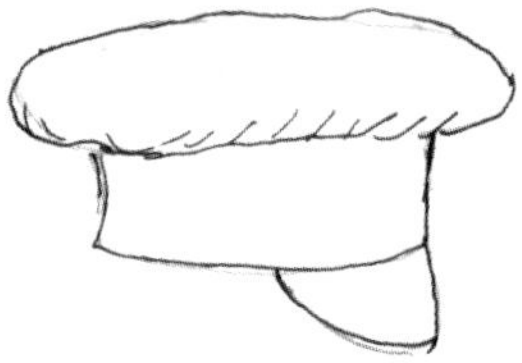

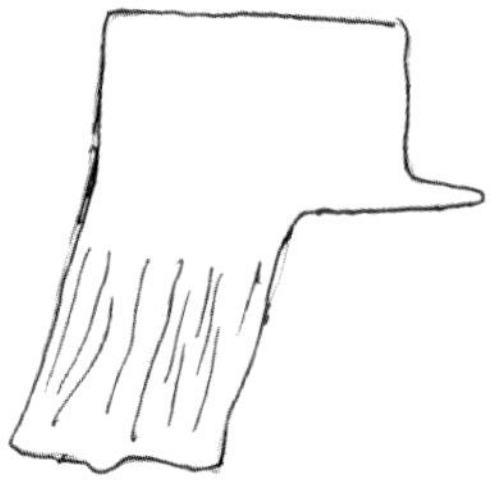

(Left) British India Forage Caps c. 1870

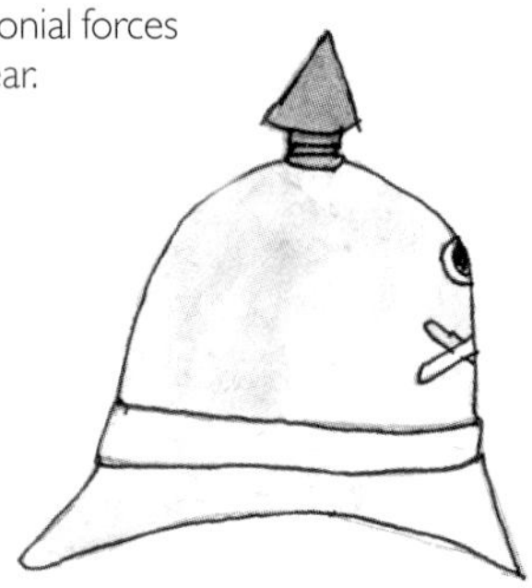

(Above) Portugese East Africa c1911

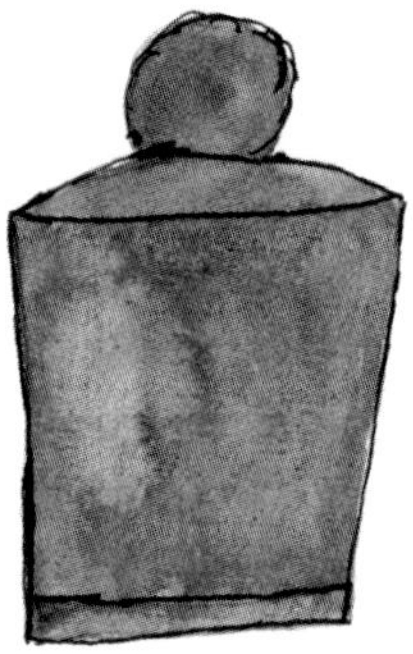

(Left) Rare Shako c 1850

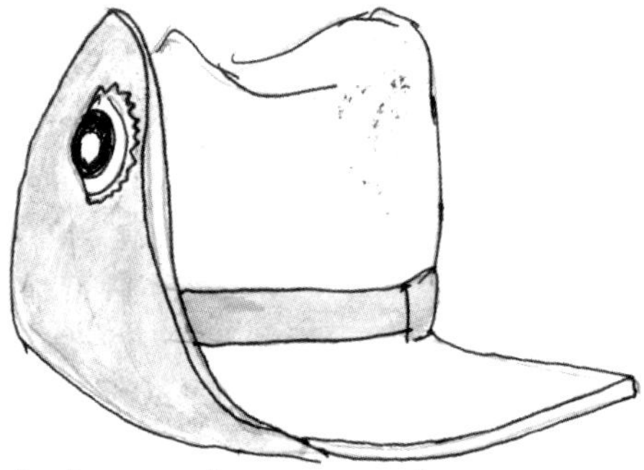

(Above) Prussian East Africa c 1900

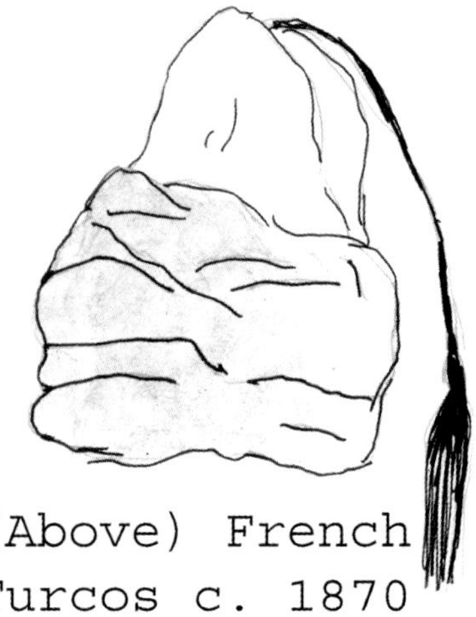

(Above) French Turcos c. 1870

Conial/Ethnographical Military Headdress

Chechia French N. African troops c. 1855

Chechia French N. African Troops c. 1842

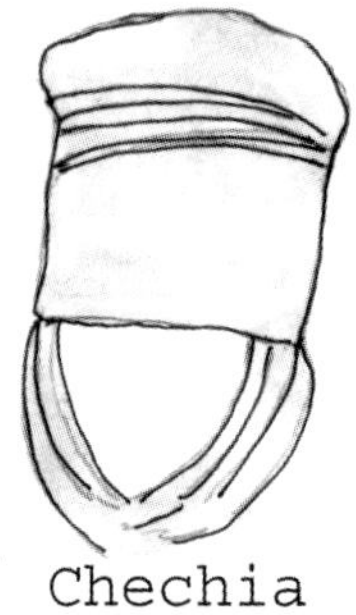

Chechia French Turcos c. 1870

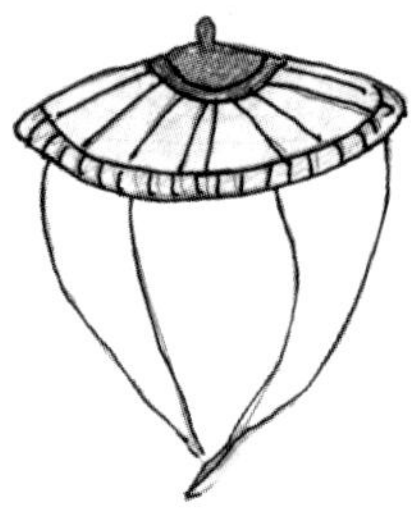

Sallacco French SE Asia 19th century

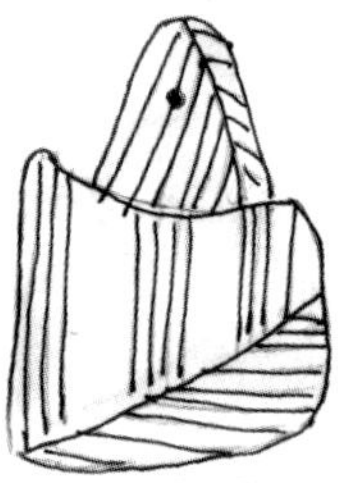

Kulah India c. 1880

Askari Prussian E. Africa c 1914

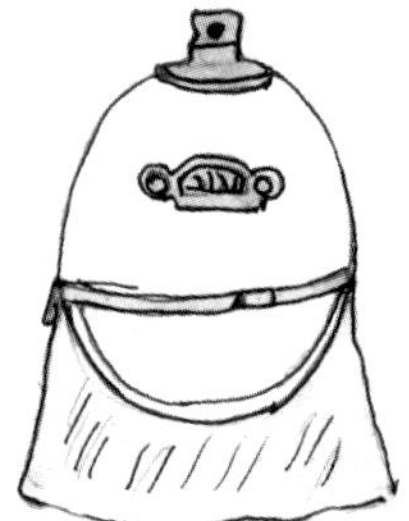

Prussian Mesopotamia c1914

Slouch Hat US Phillipines c.1890

A period illustration from the Spanish American war depicting U.S. troops wearing slouch hats. (Private collection - Malta) Photo by M Bonnici

A 19th century photograph showing a Maltese soldier wearing a colonial fatigue cap. (Private collection - Malta) Photo by M Bonnici

Notes:

[1] This leading British company continued to produce Pith Helmets well into this century. Labels of Hawkes have been observed by the author on Victorian Helmets as well as mid-twentieth century examples. Vide Kent Sales catolgue no. 207 item no. 144 as well as Wallis and Wallis Catalogues 1980-1992.

[2] The adverts appeared in various issues of The Malta Observer of 1858.

[3] The resilience of these helmets is amazing. Although they were made of fragile materials, some period pieces managed to survive to our times.

[4] A above-quoted Helmet sold by Kent Sales had a label which stated that the Hawkes Company had a retail office at Calcutta and Simla. Similar retail outlets must have been spread across all the British colonies.

[5] Vide the Illustrated London News April 25 1903;

[6] Hawkes was not the only company to produce Pith Helmets. Scott & Co of 1 Old Bond Street also was another Victorian manufacturer (Vide Kent Sales Catalogue 207)

[7] This explains the reason why it was often referred to as Solar Helmet or Tropical Helmet;

[8] The 'rival' of the Pith helmet was the Shako made out of leather and/or fabric. Like the Pith helmet it offered no protection from bullet and shrapnel strike. The Pith helmet was more appropriate for the colonies than the Shako because the Pith helmet was lighter and it could also withstand exposure to the sun.

[9] The Pith Helmet consisted of a canvas covered cork shell which was less than 5 millimetres thick.

[10] Around three main types of Pith Helmets were created: The Wolesely Pattern, the South African Pattern and the Aden Pattern.

[11] Pith helmet were often referred to as Overseas or Colonial Helmets;

[12] This is another name for the Blue Cloth Helmet;

[13] Officer helmets of this quality appear on the heads of British Officers in an 1878 engraving in the Illustrated London News published in July 6, 1878. Similar helmets were described and sold by Wallis and Wallis.

[14] This emerges from a Maltese desertion law suit Regina vs. Giovanni Camilleri per Baron A. La Chapelle (3rd of December 1897).

[15] Vide contemporary photos reproduced in John Mizzi's Gallipoli: The Malta Connection. When Malta was used as a staging post for this spectacular battle very large amounts of Pith helmets appear to have been shipped to Malta because mint and unissued period pieces appeared on the collectors' market at Malta a few years ago.
[16] Rankin op cit p.109;
[17] Vide Mollo/Macgregor/Smith/Chapelle: World Army Uniforms since 1939;
[18] Soldiers were not the only persons to wear this helmet, the Bombay bowler was also worn by civilians.
[19] These were shaped like Jockey hats;
[20] Payne. H.: the Charge of the Queen's Bays against the Mutinees at Luknow;
[21] Farwell, Byron: Queen Victoria's little wars - Plate 6
[22] Farwell Byron: Op. Cit.- Plate 7
[23] Farwell Byron: Op. Cit.- Plate 8
[24] Farwell Byron: Op. Cit.- Plate 9
[25] Farwell Byron: Op. Cit.- Plate 10
[26] Woodville Caton: All that was left of them
[27] Mizzi, Lawrence: Gallipoli - the Malta Connection pg. 94
[28] The Royal Marines and the Malta Mounted Police still wear Pith helmets on Ceremonial Occasions.
[29] Chasseurs D' Afrique;
[30] Regimentals Lot 29118
[31] Regimentals Lot 25058
[32] Regimentals Lot 26001
[33] Regimentals Lots 29120, 29321, 29121

10

The Age of Steel

From the Western Front to the Russo-Afghan War

"Before I knew, the Dawn was on the road
Far from my side, so silently he went,
Catching his golden helmet as he ran,
And hast'ning on along the dun straight way,
Where old men's sabots now began to clack
And withered women, knitting, led their cows,
On, on to call the men of Kitchener
Down to their coasts,—I shouting after him...."

John Finley
The Road to Dieppe
1917

A c. 1914 photograph that illustrates an Austrian Lancer wearing an anachronistic shapka of Napoleonic inspiration (*The War Illustrated*). (Private collection - Malta)† Photo by M Bonnici

The Belgian army used headgear that incorporated Franco-Prussian elements. This colorful early First World War Belgian Belmutze is of Prussian inspiration. *(Courtesy of Johan Somers)*

The early First World War saw the use of anachronistic, colourful headgear, such as this Kepi. (Private collection - Malta) Photo by M Bonnici

Top view of the Kepi; note the Hungarian knot on the top. Photo M Bonnici

In the opening days of World War One, serious anti-ballistic metal helmets were not issued to the troops. Soldiers were despatched to the battlefields of Tannenburg, Mons, and Liege wearing totally anachronistic headgear that traced its origins back to the Crimean war. The British expeditionary force was equipped with soft floppy caps, the French army with colourful Kepis, and the germans wore their inadequate leather Picklehaubes.[1] The all-encompassing aim of mobility on the battlefield induced military tacticians to denude their infantrymen from the life-saving metal helmet and, predictably, head injuries caused several futile deaths.[2]

By 1915 the top slots of all the military establishments decided that something had to be done to arrest the enormous amount of casualties falling to gory head injuries. After a few months of experimentation with a number of bizarre helmet designs, the Empires and nations that fought the Great War developed their distinct helmet patterns, mass produced them in World War One, and issued them to the troops in enormous numbers. Although most of these helmet patterns were based on late medieval helmets, the helmet patterns influenced the evolution of metal headgear throughout the 20th century.

Inner view of the Kepi. Photo M Bonnici

A British First World War Period Enlisted man's khaki peaked cap. *(Courtesy of Johan Somers)*

A clipping from *The War Illustrated* depicting a Turkish soldier wearing an unobtainable characteristic form of headgear worn by Turkish troops in World War One. (Private collection - Malta) Photo by M Bonnici

10.1 Experiments in steel

"When they stopped at a shell hole and washed the mud from their faces, the English boy, with his helmet off and the weather stains removed, showed a countenance of adolescent freshness, almost girlish; cheeks like pink apples, yellow curls above his forehead, long, soft lashes."

Will Carter
One of Ours
1922

The first persons to realise that some form of headgear that could offer protection against bullets and shrapnel was necessary were the soldiers themselves. This serious shortage gave rise to amateurish attempts in ballistic helmet production.

In 1915 the *Armee Abteilung Gaede* acquired 1,500 queer helmets that consisted of a sturdy heavy bullet-proof bowl and a Norman style nasal bar. The German war ministry did not approve these queer helmets.[3]

Italy issued an experimental helmet in the same year. It resembled the upper portion of a welding mask and was known as the Farina helmet.[4] The Belgian Army also used this helmet, but the threatened Belgians were creative enough to introduce their own experimental helmet patterns in the late First World War. In 1917 Belgium commissioned the Sankey foundry in Wolverhampton one thousand sallet style helmets fitted with a visor of medieval inspiration. The latter helmet became known as '*Reine Elisabeth*,' and was used in combat. However, most experimental helmets remained non-regulation issue prototypes and were rendered obsolete[5] by the introduction of the standardised mass produced regulation issue steel helmets.[6] The United States carried out extensive experiments with prototype helmets. One of these helmets was inspired by the 16th century armet, but none of these grotesque helmets was ever issued *en masse*.[7]

10.2 A Helmet for all Nations

"...many men die in the trenches from the effects of head wounds caused by snipers. The helmet is intended to minimise these casualties."

The War Illustrated
21 August 1915

The first nation that introduced the Metal Helmet was the nation that had war at its doorstep. The French republic issued its first batch of steel helmets in summer 1915. It appears that, at first, this enlightened measure was deemed to be anachronistic. The British press commented that it was a return to the middle ages, and compared the crested helmet to the helmets in Napoleonic style paintings by the celebrated painter Detaille;[8] but the history of the helmet introduced by General Adrian was much more articulate[9] than that.

France had been toying with the idea of an infantry helmet since 1836, when a prototype leather helmet was created.[10] A much more elaborate prototype appeared in 1852. It was a leather helmet of Picklehaube form, but it retained some distinctively French features, such as the v-shaped brim, the short neck flange, and the banded helmet plate. It was the only French helmet to be surmounted by an eagle. It was too expensive to produce, and it was not issued as a regulation helmet. After experimenting with the use of Blue Cloth helmets, the French military issued another trial helmet; the Model 1901 Andre', which influenced the evolution of a number of trial helmets with flared brim and raupenhelm form. In 1912 the painter Detaille created a trial casque based on this helmet. This helmet included some innovations on the Andre' models. It had a dished peak of Prussian inspiration and, more importantly, a brass comb as worn by the firemen of the period. This helmet appears to have seen some occasional use. It also received advanced publicity in the press, but was not issued as a regulation helmet.[11] The French army had to wait for the First World War to obtain its first metal helmet.

In the early First World War the French *Poilu* had, like his enemy, the hated *boche*, and improvised his own helmet. This odd helmet consisted of *cervelliere* (hemispherical metal skull caps) that were worn under the Kepis and other crude contrivances. In July 1915 the French Army finally issued the Adrian helmet. By December 1915 three million Adrian helmets were distributed to the troops.[12] The M.1915 was embellished with a range of helmet plates. Light Infantry used a bugle plate reminiscent of the first empire, engineers used a Greek panopoly helmet plate, medics and paramedics used a helmet plate that represented a snake intertwined to a column, the artillery used a crossed cannons device, and zouaves used a half moon device, while the rest of the infantry used a grenade device.

The Adrian M.15[13] had a rather complex construction. It was built from no less than 5 pieces: the comb; the front brim; the back brim; the skull; and the helmet plate.[14] This unpractical method of construction was not the only defect the Adrian suffered. The Adrian was made from inadequate light gauge steel.[15] This notwithstanding, the success of the Adrian helmet knew no bounds, and the nations that imported the 1915 model Adrian helmets from France were many: Russia; Estonia;

Back to the Middle Ages: Armour for Infantry

British soldiers with some of their French friends who are wearing the new trench helmet. It is made of specially hardened steel, after drawings by Detaille, the celebrated French battle-painter.—The helmet has a rim which acts as a protection for the eyes.

As in the days of Agincourt. French infantry, wearing the shrapnel-proof helmet, taking cover behind a hedge. The British Government has purchased a number of these head protectors for trial use in the trenches. Inset: French helmeted sentry. Many men die in the trenches from the effects of head wounds caused by snipers. The helmet is intended to minimise these casualties.

A clipping from the 21st August 1915 edition of *The War Illustrated*. It announces the issue of the M 1915 Adrian helmet. (Private collection - Malta) Photo by M Bonnici

A model 1915 Adrian helmet issued to the Romanian army. *(Courtesy of Johan Somers)*

Cechoslovakia; the United States of America; Italy; Thailand; Serbia; Rumania; Poland; and Belgium.

A number of countries created their own Adrian helmets. After using the French Adrian in the first year of the war the Italian iron smelting works created an improved Adrian helmet that was made from sturdier gauge steel and fewer pieces.

Sweden created an experimental Adrian helmet in the 1920s, but this helmet was subsequently discarded.

The War Illustrated of August 1915 recounts that the British Army acquired some Adrian helmets from the French ally 'for trial use,' but the trials may have had an adverse outcome, because a British pattern steel helmet was issued in 1916.[16] The British Empire wore the legendary Brodie Model.

A Model 1915-1916 Model Italian Adrian helmet. In the early First World War Italy imported its helmets from France, but as the war progressed the Italians created their own variant of the Adrian helmet. (Private collection - Malta) Photo by M Bonnici

Close up view of the WW I Italian helmet. Applique' badges were seldom worn by the Italians, and regimental badges were often stenciled on the helmets. (Private collection - Malta)†Photo by M Bonnici

The French Adrian helmet was a source of inspiration for many helmets, this rare Swedish helmet included. *(Courtesy of Stefan Anbratt)*

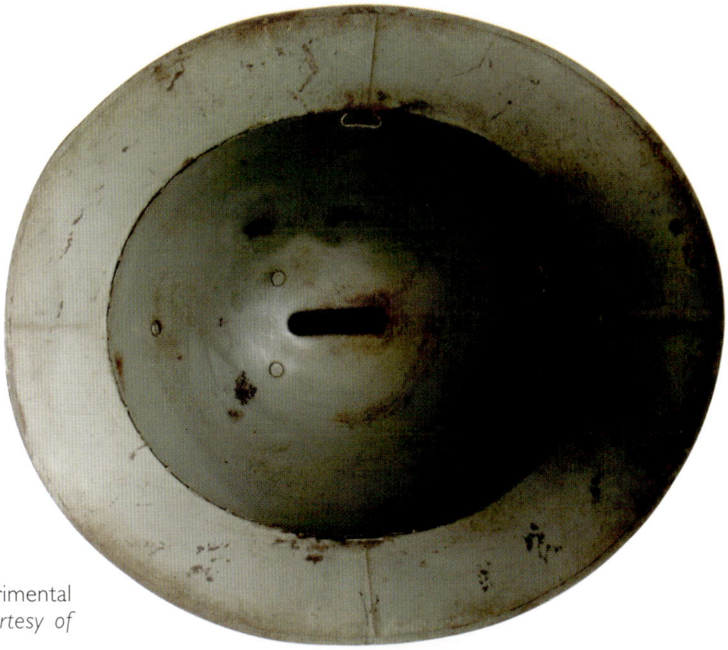

The inside view of the experimental Swedish Adrian helmet. *(Courtesy of Stefan Anbratt)*

An early World War One Brodie pattern British helmet in its original green finish. (Private collection - Malta)† Photo by M Bonnici

Inside view of the Brodie helmet. Photo M Bonnici

The U.S. version of the Brodie helmet. (Private collection - Malta)†Photo by M Bonnici

The Brodie model was inspired by the Medieval Siege kettle hat of the days of the Black Prince. It looked like a ravioli.[17] When the United States of America joined the war it issued its Doughboys facsimiles of the Brodie Helmets. The U.S. counterpart of the Brodie helmet was named the USM1917.

Portugal—Britain's oldest ally—used Brodie pattern helmets, but it also used a special mild steel helmet (made in Birmingham) with a distinctive corrugated bowl, denominated M.16.[18]

During the Great war, the Germans issued three different helmets to their troops. The first to appear was the solid M.1916 inspired from the Gothic Sallet of the Golden days of Chivalry. This helmet went into active service in December 1915. It was improved in 1917/1918[19] when its suspension system was modernised. A third model (known as the cavalry pattern) was produced in the last year of the War. It had an unusual ear cut-out that enabled telephone operators to listen to their equipment.[20]

The Austro-Hungarian Empire issued its troops a facsimile of the German M.16,[21] christened the M.1917. In the early part[22]

A Portugese Model 1916 helmet. *(Courtesy of Johan Somers)*

A German model M 1916 Stahlem. This helmet could be fitted with a brow reinforcement. (Private collection - Malta) Photo by M Bonnici

Close up view of the helmet's liner. A name was scrolled on one of the pads. Photo M Bonnici

Front view of an Austrian M 16/17 with the hand painted insignia of a stormtrooper unit. (Private collection - Malta) Photo by M Bonnici

Side view of an Austrian Model 1916 Stahlem. Photo M Bonnici

Close up view of the Austrian helmet's cloth strap. Photo M Bonnici

of the War the Austrians developed what could have been the source of inspiration for the German M.16, the famous Berndorfer helmet. The Berndorfer helmet had a similar shape to the M.16, but had no projected brim. The main feature of this helmet was the ventilation hole at the top of the helmet. This helmet fell out of use as the war progressed.[23]

Some of the steel helmets of the First World War were re-issued in the Second World War, and served as a source of inspiration for further helmet patterns. This was the case with the French M.15, the German M.16 and M.17/18, and even the Brodie helmets. However, the post war witnessed the rise of a new breed of helmets. Most of these were styled on an American design patented in 1940.[24] Some the latter were made of hard plastics and gave rise to a new era in the history of headgear.

The post Second World War period witnessed the rise of a breed of U.S. M1 helmets, such as this hard plastic helmet worn by Saddam's infantry in the first gulf war. (Private collection - Malta)† Photo by M Bonnici

An Austrian officer's fatigue cap. The First World War witnessed the adoption of practical articles of headgear, such as this type of cap c. 1918. *(Courtesy of Johan Somers)*

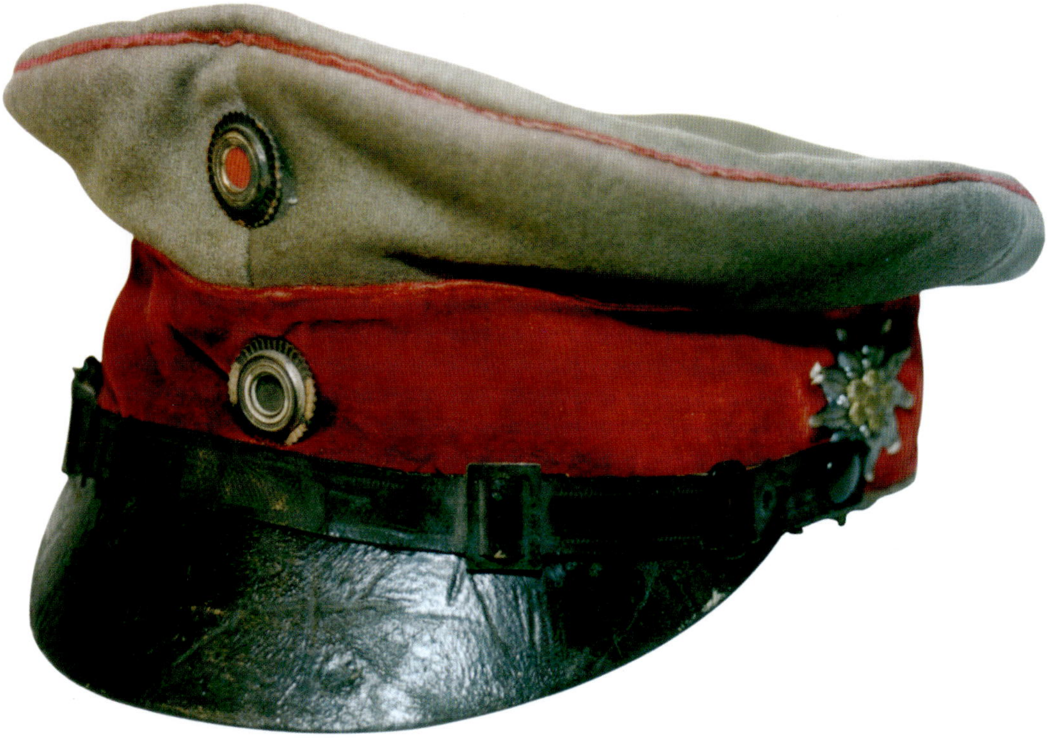

A rare Prussian officer's peaked cap of the Alpenkorps c. 1917. *(Courtesy of Johan Somers)*

A Prussian officer's peaked cap c. 1916. *(Courtesy of Johan Somers)*

10.3 Collecting

A large quantity of First World War steel helmets are regularly featured on internet auction listings, and dealers of militaria are always well stocked with these collectibles. Collectors have a wide range to choose from, including the unissued minty pieces sought for by the wealthy collectors, as well as the affordable, but equally interesting rusty battle relics. Relics (sans liner and rivets) may sell for as little as USD 100, while rare experimental and prototype helmets fetch very high prices. In 1997 a collector offered for sale an exceptional visored French M.15 helmet at around USD 750.[25] Around two years earlier a British dealer offered a fine German M.18 helmet, as ordered by Turkey, at around USD 2,300.[26] Mass produced helmets sell for significantly lower prices.

In September 2003 a British dealer offered an Austrian M.17 in mint condition at USD 360[27] and a similar specimen in good condition at USD 216.[28] The quasi-identical average German M.16 sells within a slightly lower price range. In fact, the same dealer offered a German M.16 in good condition and with Hindenburg scroll at USD 266.[29] Camouflaged M.16 helmets sell for higher prices. In June 2003 another British dealer offered a rare, very small size M.16 helmet retaining 60% of its block camouflage finish at USD 536,[30] and a camouflaged M.17 helmet in excellent condition at USD 635.[31] French helmets appear to be more common, and even examples with scarce helmet plates sell within a more or less modest price range. In June 2003 a French M.15 with medical badge was offered for sale at USD 198,[32] while in September another dealer offered an equally scarce colonial Adrian helmet with half moon plate at USD 208.[33] The latter dealer offered a run of the mill infantry French M.15 in good condition at USD 122,[34] and a slightly worn French M.15 with engineer's badge at USD 108.[35]

British Brodie helmets are becoming very hard to find, and are consequently becoming relatively expensive. The early models are the most sought after specimens. In October 2003 a dealer offered an early First World War helmet in a quasi-immaculate condition at USD 349,[36] a private purchase rimless Brodie at USD 449,[37] and a rare Australian unit marked helmet at USD 950.[38] The same dealer offered a couple of late war Brodies in fair condition at USD 129[39] and USD 99.[40] U.S. M.17 Brodie helmets sell within the latter price range. In October 2003 a dealer offered a helmet with insignia at USD 95.[41]

The author does not know of the production of fake metal[42] British and American First World War helmets, but occasionally Second World War helmets are passed off as First World War helmets. British Second World War helmets and Belgian M.1949 helmets are frequently confused with their much rarer First World War Brodie equivalents. The hallmarks of First World War helmets include the lining system, the small mounting rivet at the top of the helmet, and the squattish helmet bowl. The author did not come across fake German metal helmets, but he did notice some concerning original German helmets altered[43] to appear like the extremely rare helmets produced for Turkish troops in 1918. Original helmets painted with replica camouflage paint are equally concerning. It is frequently very difficult to identify these forgeries, but a trace of rust bubbling underneath the camouflage finish is always a bad sign. The author recently learned that original relic German M.18s and German M.16s are being fitted with lower quality post-First World War Swedish helmet liners and sold as complete helmets.

The War Illustrated, 21st August, 1915.

Regd. as Newspaper & for Canadian Magazine

"Onward!" Be Our Watchword, by A. G. Hales

The War Illustrated

2d. Weekly

Vol. 3 The Second Year! Civilisation's Seven Allies still shoulder to shoulder against the Hun No. 53

The cover of *The War Illustrated*, 21 8 15. It illustrates allied soldiers wearing their characteristic headgear. (Private collection - Malta) Photo by M Bonnici

Notes:

[1] Wilkinson op.cit p. 37;
[2] Becker Edwards op. cit 1-10;
[3] Marzetti Paolo op. cit p. 89;
[4] Various patterns of this helmet were issued to the troops;
[5] The Belgian M.17 Reine Elizebeth is the obvious exception, this helmet appeared following the introduction of the Adrian Helmet in 1915;
[6] Marzetti op.cit p.27;
[7] Marzetti op.cit pp.277-281;
[8] The War Illustrated, 21st August 1915;
[9] The War Illustrated, 21st August 1915;
[10] This helmet was later issued to firemen in lieu of the infantry;
[11] Tavard op.cit pp.99-101;
[12] Author Unknown: The Adrian Pattern M1915 Steel Helmet published in the Armourer Magazine Issue 5;
[13] After the First War the French created a second Adrian helmet, the Model 1926.. The 1926 Model was made out of fewer pieces and forged in heavier gauge steel. The M.26 is much more common than the M.15. The M. 15 can be distinguished from its later counterpart from its distinctive ring round the crown.
[14] In some specimens the Adrian was fitted with brw reinforcements and visors;
[15] Tavard op.cit pp.104-108;
[16] This helmet had been patented on August 16th 1915; Vide Marzetti op.cit p.211;
[17] Two models were used in the war. The early model (Mark I) featured no rim while the later model (Mark II) was fitted with a steel rim which softened the sharp edges of the helmet
[18] Soares Branco op.cit pp.176-177;
[19] A modified example of the German M. 1917/18 was ordered for the Turkish troops. The Turkish M. 1918 did not feature a peak.
[20] Taylor Andrew: The German M1916 pattern helmet pp.28-30;
[21] The Hallmark of the Austrian Helmet was the cloth strap, the rearword lining rivet, and the brownish colour.
[22] Prior to the introduction of the M.16;
[23] Marzetti op.cit pp.20-21;
[24] Marzetti op.cit p.284;
[25] The Armourer Magazine Issue 19 p.6;
[26] Regimentals Catalogue No. 39 item 989;
[27] Alan Beadle Limited Item 1713;
[28] Alan Beadle Limited Item 0013;
[29] Alan Beadle Limited Item 0623;
[30] Regimentals Item 25092;
[31] Regimentals Item 25962;
[32] Regimentals Item 26697;
[33] Alan Beadle Ltd 0558;
[34] Alan Beadle Ltd 0946;
[35] Alan Beadle Ltd 0561;
[36] Great War Militaria GB10004;
[37] Great War Militaria GB6006;
[38] Great War Militaria GB6008;
[39] Great War Militaria GB5813;
[40] Great War Militaria GB5893;
[41] No Man's Land Militaria Item#302;
[42] Electrotype facsimile German World War One helmets are in circulation but the latter replicas are made of fibreglass.
[43] With the peak removed;

11

The Near and Far East

A Note on Oriental Helmets

"Unlike the European Student of armour, the Oriental expert, as a rule, is influenced but little by a gold hilt enamelled and set with jewels or any other valuable enrichments to learn the advantage that may appear on a weapon...."

Guy Francis Laking
1913

Western Europe's interest in Oriental Arms and Armour dates back to the second half of the 19th century, when famous collectors, such as Lord Hertford, started to hoard massive collections of Indo-Persian and Japanese armour.[1] Collections of Oriental arms and armour sprouted in England,[2] the United States of America,[3] and Italy.[4] Travellers acquired Japanese Samurai Helmets as a souvenir from a distant land, while

A 19th century newspaper clipping that illustrates Persian warriors wearing Kulah Khudds on the Caucasus. (Private collection - Malta) Photo by M Bonnici

A 19th century portait of the Rajah of Patala. He wears a Kulah Khudd. (Private collection - Malta) Photo by M Bonnici

veterans of the Indian wars brought home Indo-Persian helmets as spoils of war. Most of the Oriental Helmets on the modern collectors' market derive from these enormous Victorian collections. Unfortunately, this legacy seems to be reflected in Western Europe's approach to these valuable artefacts. The average collector's knowledge of Oriental Helmets is rather low, because these helmets are rarely appreciated for their history, and are simply looked upon as mere curiosities.

The Islamic helmet

The origins of the Islamic helmet may date back to the Pyllos helmets worn in antiquity. The celebrated Nimrud bas reliefs depict soldiers wearing tall conical helmets with hinged cheek pieces similar to those worn in the Ottoman Empire[5] from the 14th century to the 16th century.[6] In the fifteenth century this tall, conical helmet known as the Shishak developed in two versions. The first variety (known as the Shishak) was fitted

A rare indo-persian kulah khudd of Melon form from the late 18th century (ex. Wallis and Wallis published in catalogue 414, plate 17). (Private collection - Malta) Photo by M Bonnici

Side view of the Kulah Kudd. Photo M Bonnici

Close up of the helmet's nasal guard and twin plume holders. Photo M Bonnici

Rear view of the Kulah Kudd. Photo M Bonnici

with lamellar cheek pieces and brims, and later influenced the development of the Zischagge. The second (known as the Kulah Khudd) was fitted with a mail tippet in lieu of lames, and was occasionally surmounted with a rectangular spike. Both versions were worn underneath or over turbans and became known as turban helmets. The Shishak was particularly popular in the Ottoman Empire, while the Kulah Khudd was very popular in India and Persia. Both types of Islamic helmet were worn in Egypt,[7] North Africa, Arabia,[8] and Syria. The Shishak fell in disuse in the 17th century with the decline of armour, but the Kulah Khudd remained in use in India and Persia up to the nineteenth century.[9]

Most of the authentic Shishaks in circulation (with the exception of a polished metal helmet worn by the Egyptian Horse Guards in the early 19th century) are pre-1700, but most of the authentic Kulah Khudds in circulation date from the 19th century. It is frequently very difficult to distinguish pre-1600 Kulah Khudds from later Kulah Khudds because information on the subject is very hard to find.[10] John Wallace published a highly informative article on the subject in 1971.[11] Wallace quotes Stocklein, who sustains that pre-1600 helmets were forged in one piece with the spike, while later helmets were made in two pieces. The mail used is an equally important dating feature; post 1700 helmets were frequently fitted with butted, as opposed to riveted, mail tippets. Most Kulah Khudds were embellished with gilding, etching, or silvering. Most Kulah Khudds were also fitted plume holders, while some odd rare versions were fitted with a pair of iron horns.[12]

The Kulah Khudd was not the only helmet worn in Persia and India. Use was also made of very odd helmets made entirely of mail, or of a combination of mail and plate.[13] The latter helmet was known as a Kulah Zirah and is a scarce helmet.[14]

The rarity of pre-1700 western European armour in the past few years appears to have given rise to a surge in the popularity of most forms of Islamic armour of Saracenic inspiration. In 1994 a British dealer offered a finely chased Indo-Persian helmet at approximately USD 1,000.[15] In June 2003 an antique Middle Eastern Kulah Khudd was offered for sale at USD 1,675[16] but remained unsold. In October 2003 an internet dealer offered a complete 18th century Indo Persian helmet made from riveted mail at around USD 1,350.[17] Another internet dealer who appears to specialise exclusively in Oriental arms and armour offered an early Indian 19th century chiselled Kulah Khudd at USD 1150,[18] a 17th century Kulah Zirah at USD 950,[19] and a fluted North Indian Kulah Khudd at USD 1,450.[20]

A 19th century etched kulah khudd. *(Courtesy of Fondazzjoni Patrimonju Malti)*

Fake Kulah Khudds are presently being produced in India, but these replicas are of a much lower quality than the originals.

The Kabuto and the Jingasa

The Kabuto is one of the most evocative helmets in history. It was the helmet of the Samurai, and in Japan Kabutos are conserved with a quasi-religious form of attention. In Japan the Kabuto appears to have been the subject of numerous publications, but unfortunately very few publications dedicated exclusively to the Kabuto have been published in a European language. This dearth of information on this highly specialised area of collecting has given rise to various misconceptions.

It appears that the Kabuto is a helmet of an ancient inspiration. The shape of the Kabuto was noticed by Turnbull

An excavated Kabuto of toppai (peach) form, probably late Muromachi period. (Private collection - Malta) Photo by M Bonnici

A russetted 18th century Kabuto fitted with Maidete in the form of a demon. *(Courtesy of Mukashiya K.K.)*

Front view of the Mid Edo Kabuto. *(Courtesy of Mukashiya K.K.)*

Side view of the 18th century Kabuto fitted with Maidete in the form of a demon. *(Courtesy of Mukashiya K.K.)*

Top view of the Mid Edo Kabuto. *(Courtesy of Mukashiya K.K.)*

A close up of the Mid Edo Kabuto's hachi manza. *(Courtesy of Mukashiya K.K.)*

A view of the russetted Kabuto's shikoro.

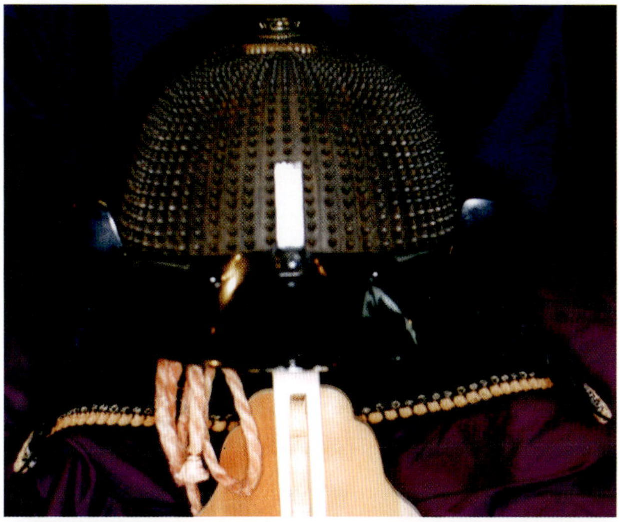

An Akoda style Kabuto that probably dates from the late Muromachi period (16th century). *(Courtesy of Mukashiya K.K.)*

Above: An Akoda style Kabuto that probably dates from the late Muromachi period (16th century). *(Courtesy of Mukashiya K.K.)*

Right: A view of the akoda Kabuto's hachi manza. *(Courtesy of Mukashiya K.K.)*

in a Japanese scroll from the *Gosannen no eki emaki in the Tokyo National Museum* dating circa 1083.[21] At this time the Kabuto had already developed most of its characteristics: the hemispherical bowl (*Hashi*) of *Spangenhelm* construction (made from overlapping plates rivetted together); the lamellar neck flange (the *shikoro*); and the projected brim (*maye-zashi*). In the 15th century the development of this helmet was heavily influenced by the taller Chinese (Manchurian) and Korean helmets. Interaction with the west in the 16th century gave rise to Kabuto of morion and burgonet inspiration. The later period saw a revival of old designs and a combination of different styles. It also appears that old *hashis* were frequently re-issued with later *shikoros*.

Most Kabutos were lacquered (painted) with layers of fine coloured veneer, giving the kabuto a† 'plastic appearance.' It is frequently very difficult to date Kabutos, but it appears that the smaller the number of lames in the bowl (the *Hashi*), the older the helmet. The shape of the *Hashi* is similarly another dating feature.[22] Most of the helmets of a western european inspiration date back to the 16th, 17th, and 18th centuries. Most of the Kabutos that appear for sale are from the Edo Period (c.1603-1867[23]), but occasionally a rare early example from the Muromachi period (c.1333-1573[24]) does appear for sale, and it is frequently unnoticed by most western European collectors.

The second form of Japanese helmet is the flat plate shaped Jingasa used by foot soldiers and retainers. The Jingasa was a military helmet and a helmet for dress use. Some jingasas were made from a single metal plate, while others were made from a series of overlapping plates. Parade and ceremonial Jingasas were made of wood. They were often finely painted with futuristic motifs and bright colours. Jingasas were fitted with heavy padded liners.[25]

Kabutos and Jingasas are in demand, and this demand has prompted the creation of replicas. Unfortunately, some high quality replicas have penetrated the market, and these are often mistaken for the real thing, on account of the fact that they

An 18th century Kabuto known as a suji-Kabuto laquered in black. *(Courtes of Mukashiya K.K.)*

View of the suji-Kabuto's shikoro. *(Courtesy of Mukashiya K.K.)*

Front view of the Suji-Kabuto laquered in black. *(Courtesy of Mukashiya K.K.)*

A view of the Suji Kabuto's hachi manza. (Courtesy of Mukashiya K.K.)

A view of the Suji Kabuto's back. *(Courtesy of Mukashiya K.K.)*

A Kabuto from the mid-Edo-Period (c. 1750). *(Courtesy of Mukashiya K. K.)*

A Kabuto from the mid-Edo-Period (c. 1750). *(Courtesy of Mukashiya K. K.)*

A view of the hachi manza of the same Kabuto. *(Courtesy of Mukashiya K.K.)*

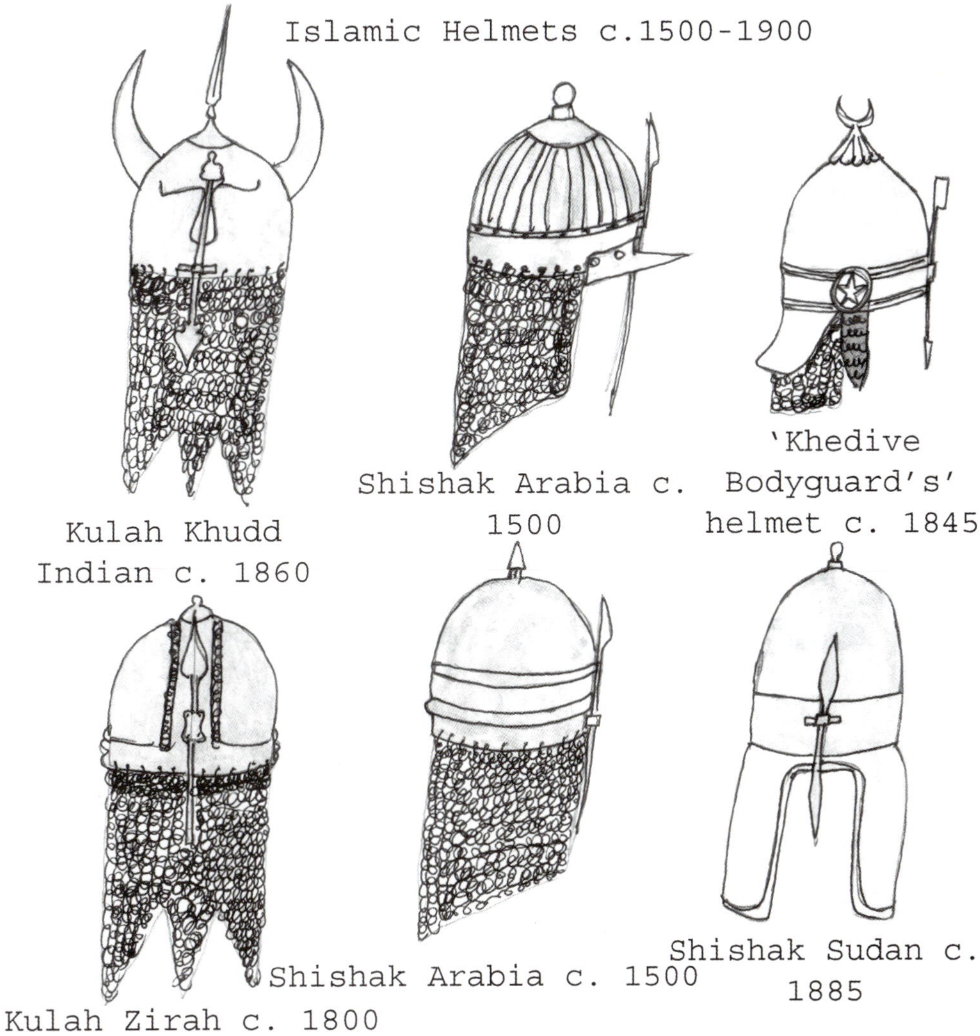

Artist's impression of a series of Islamic helmets.

were produced by artisans that used the same construction techniques used by the armourers of the Edo period. Expert advice from specialists should be sought prior to a purchase.

Kabutos sell within a vast price range. Lyle's Price Guide for 2001 listed a Kabuto *Hashi* that sold for USD 725[26] and a signed complete Kabuto dated 1539 that sold for USD 67,981.[27]

Notes:

[1] Guy Francis Laking: Wallace Collection Catalogues, Oriental Arms and Armour p.ix;
[2] Including the Wallace Collection;
[3] Including the Grosse collection;
[4] Including the Wallace Collection;
[5] Attard Robert: Islamic Helmets (article published in the Armourer Magazine Issue 27);
[6] Stone George Cameron: A Glossary of the Construction, Decoration and use of the Arms and Armour in all countries and in all times p.47;
[7] Boccia op.cit fig. 472;
[8] Boccia op.cit fig.473;
[9] Attard op.cit;
[10] Moreover it is important to keep in mind that pre-1700 were frequently refurnished and re-issued in the nineteenth century;
[11] Wallace John: Islamic Arms and Armour (published in Discovering Antiques Vol. 2 pp.548-552);
[12] These helmets were known as Devil's Heads Kulah Khudds;
[13] Gross op.cit p.398;
[14] A similar helmet was worn in Sudan;
[15] The Armourer Magazine issue 5 p.12;
[16] Ebay Item 2176652326;
[17]www.antique-arts.com: Item#AR14-M49;
[18] Ashoka Arts Ar-9;
[19] Ashoka Arts Ar-3;
[20] Ashoka Arts Ar-6;
[21] Turnbull Stephen: Samurai Warriors p.10;
[22] Stone op.cit pp.327-332
[23] www.japan.guide.com
[24] www.japan.guide.com
[25] Rankin op.cit p.41;
[26] Lyle Antique Price Guide 2001 p.88;
[27] Lyle op.cit p.91;

12

A Few Reflections on Collecting Military Headgear

"The Headgear of a people, that of an army, evolved with time, with fashion, with necessity...it reflects the trends of Èpoques, the influence of neighbouring peoples; their pomp...."

Lieutenant-Colonel Sauzey
(from the Preface to J. Margerand's Les Coiffures De l'Armee Francaise)
1909

12.1 The Philosophy of Collecting

Why do we hoard military headgear? Are we over-grown school boys playing soldiers? Do we collect soldiers' hats because we wish to relive a lost youth? Do we purchase military hats as an investment? Is collecting military headgear such a superficial pass time...?

Some collectors do consider their hats to be their toys, they do fool around with their expensive collectables on their heads when they have a drink, and do speak of their collections in a childish manner. Collecting military headgear is also a way of reliving a lost innocence, a time when our imagination still galloped at the thought of Custer at Little Big Horn or Marshall Ney at Waterloo. Occasionally, the acquisition of a good military hat may also turn out to be a valuable investment in a capricious market, but there is more to collecting military headgear than that....

Collecting military hats is tantamount to preserving history. Some of the choice pieces in the grand state collections were acquired from private collectors, and we would not have museums if we did not have dedicated collectors. Choosing to

A rare Hungarian Shako/Kepi of an officer c. 1900 in mint, untouched condition.

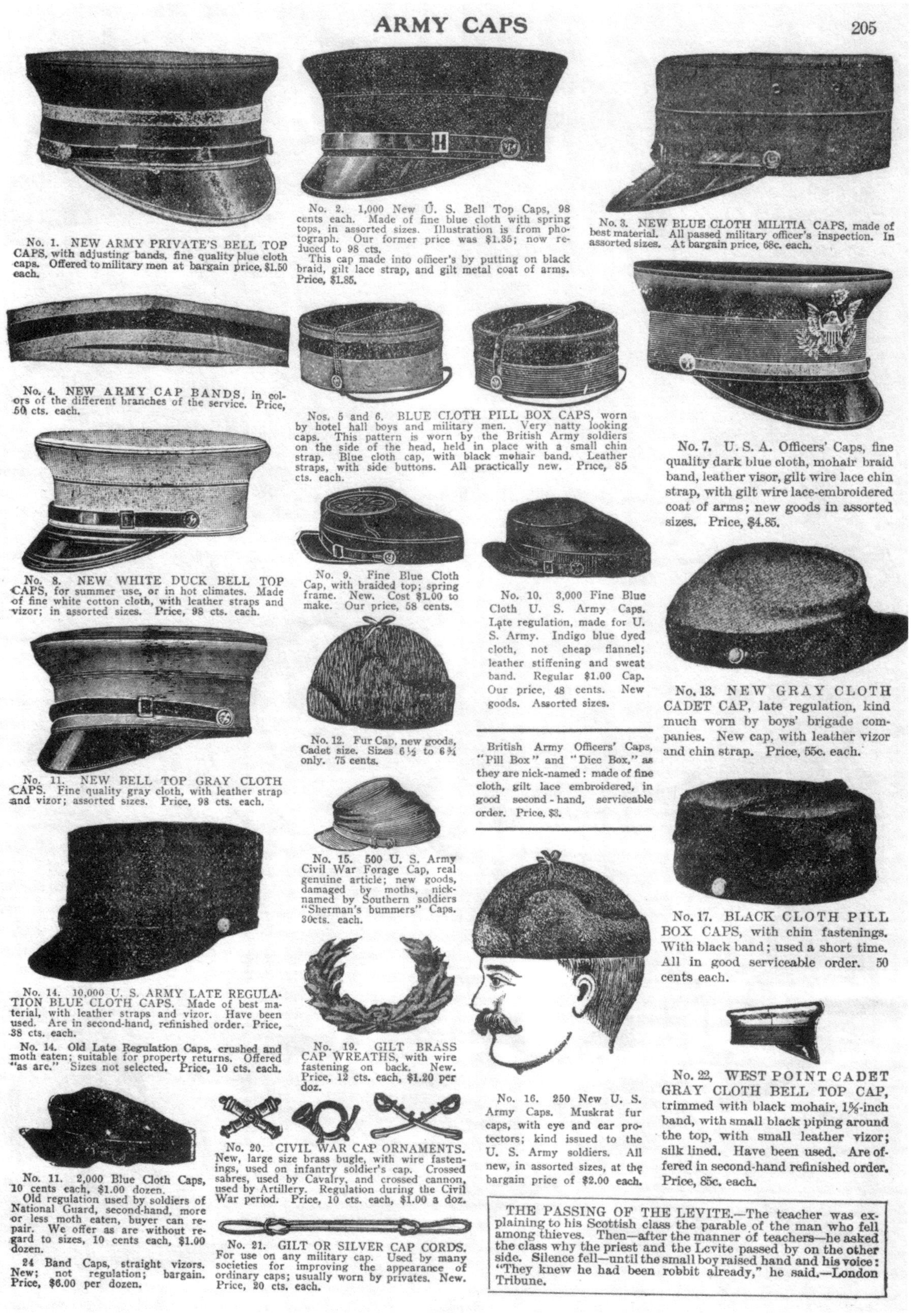

ARMY CAPS

205

No. 1. NEW ARMY PRIVATE'S BELL TOP CAPS, with adjusting bands, fine quality blue cloth caps. Offered to military men at bargain price, $1.50 each.

No. 2. 1,000 New U. S. Bell Top Caps, 98 cents each. Made of fine blue cloth with spring tops, in assorted sizes. Illustration is from photograph. Our former price was $1.35; now reduced to 98 cts.

This cap made into officer's by putting on black braid, gilt lace strap, and gilt metal coat of arms. Price, $1.85.

No. 3. NEW BLUE CLOTH MILITIA CAPS, made of best material. All passed military officer's inspection. In assorted sizes. At bargain price, 68c. each.

No. 4. NEW ARMY CAP BANDS, in colors of the different branches of the service. Price, 50 cts. each.

Nos. 5 and 6. BLUE CLOTH PILL BOX CAPS, worn by hotel hall boys and military men. Very natty looking caps. This pattern is worn by the British Army soldiers on the side of the head, held in place with a small chin strap. Blue cloth cap, with black mohair band. Leather straps, with side buttons. All practically new. Price, 85 cts. each.

No. 7. U. S. A. Officers' Caps, fine quality dark blue cloth, mohair braid band, leather visor, gilt wire lace chin strap, with gilt wire lace-embroidered coat of arms; new goods in assorted sizes. Price, $4.85.

No. 8. NEW WHITE DUCK BELL TOP CAPS, for summer use, or in hot climates. Made of fine white cotton cloth, with leather straps and vizor; in assorted sizes. Price, 98 cts. each.

No. 9. Fine Blue Cloth Cap, with braided top; spring frame. New. Cost $1.00 to make. Our price, 58 cents.

No. 10. 3,000 Fine Blue Cloth U. S. Army Caps. Late regulation, made for U. S. Army. Indigo blue dyed cloth, not cheap flannel; leather stiffening and sweat band. Regular $1.00 Cap. Our price, 48 cents. New goods. Assorted sizes.

No. 12. Fur Cap, new goods, Cadet size. Sizes 6½ to 6¾ only. 75 cents.

British Army Officers' Caps, "Pill Box" and "Dice Box," as they are nick-named: made of fine cloth, gilt lace embroidered, in good second-hand, serviceable order. Price, $3.

No. 13. NEW GRAY CLOTH CADET CAP, late regulation, kind much worn by boys' brigade companies. New cap, with leather vizor and chin strap. Price, 55c. each.

No. 11. NEW BELL TOP GRAY CLOTH CAPS. Fine quality gray cloth, with leather strap and vizor; assorted sizes. Price, 98 cts. each.

No. 15. 500 U. S. Army Civil War Forage Cap, real genuine article; new goods, damaged by moths, nicknamed by Southern soldiers "Sherman's bummers" Caps. 30cts. each.

No. 17. BLACK CLOTH PILL BOX CAPS, with chin fastenings. With black band; used a short time. All in good serviceable order. 50 cents each.

No. 14. 10,000 U. S. ARMY LATE REGULATION BLUE CLOTH CAPS. Made of best material, with leather straps and vizor. Have been used. Are in second-hand, refinished order. Price, 38 cts. each.

No. 14. Old Late Regulation Caps, crushed and moth eaten; suitable for property returns. Offered "as are." Sizes not selected. Price, 10 cts. each.

No. 19. GILT BRASS CAP WREATHS, with wire fastening on back. New. Price, 12 cts. each, $1.20 per doz.

No. 22, WEST POINT CADET GRAY CLOTH BELL TOP CAP, trimmed with black mohair, 1⅝-inch band, with small black piping around the top, with small leather vizor; silk lined. Have been used. Are offered in second-hand refinished order. Price, 85c. each.

No. 16. 250 New U. S. Army Caps. Muskrat fur caps, with eye and ear protectors; kind issued to the U. S. Army soldiers. All new, in assorted sizes, at the bargain price of $2.00 each.

No. 11. 2,000 Blue Cloth Caps, 10 cents each, $1.00 dozen.

Old regulation used by soldiers of National Guard, second-hand, more or less moth eaten, buyer can repair. We offer as are without regard to sizes, 10 cents each, $1.00 dozen.

24 Band Caps, straight vizors. New; not regulation; bargain. Price, $6.00 per dozen.

No. 20. CIVIL WAR CAP ORNAMENTS. New, large size brass bugle, with wire fastenings, used on infantry soldier's cap. Crossed sabres, used by Cavalry, and crossed cannon, used by Artillery. Regulation during the Civil War period. Price, 10 cts. each, $1.00 a doz.

No. 21. GILT OR SILVER CAP CORDS. For use on any military cap. Used by many societies for improving the appearance of ordinary caps; usually worn by privates. New. Price, 20 cts. each.

THE PASSING OF THE LEVITE.—The teacher was explaining to his Scottish class the parable of the man who fell among thieves. Then—after the manner of teachers—he asked the class why the priest and the Levite passed by on the other side. Silence fell—until the small boy raised hand and his voice: "They knew he had been robbit already," he said.—London Tribune.

A rare clipping from a 1927 Bannerman catalogue. (Private collection - Malta) Photo by M Bonnici

become a dedicated collector is appointing oneself as the curator of a private museum. Owning a valuable collection is not only about ostentation, conversation pieces, and prestige. Owning a well-researched collection entails a huge amount of responsibility and self-sacrifice. The discerning collector is the caretaker of history, the ambassador of forgotten glories, and the custodian of forsaken souls. Most of the soldiers who wore the hats in our collections have been forgotten, and nobody cares to pay their respects to their memory. Collecting is a way of rectifying this shortcoming. It is also a constant learning process. Collecting is about research, discovery, and a passion for history. Researchers like Joseph Margerand transformed our interest from a quasi-academic exercise. Collecting military headgear can turn out to be a very rewarding hobby. The author knows of private collections that are the envy of public museums, and of collectors who succeeded in acquiring rare pieces of significant historical importance. †Few hobbies are an equally intelligent way of spending one's free time and resources.

12.2 Some Thoughts on Crisps and Flakes

Most of the published works on military collectables seem to concur in providing one tip to their readers. Experts seem to solicit collectors to invest only in the crisp, mint pieces. They appear to advise their readers not to waste their precious money on flaky helmets and battle relics. Although I understand why items in good condition should fetch higher prices than worn examples, I strongly disagree with those who consider relics and items in fair condition as not being collectable. The acquisition of distressed helmets provided me with a unique opportunity to handle some of the most important forms of military headgear without spending a significant amount of money. Choosing to invest exclusively in museum condition examples can be rather restrictive and very costly. Buying a distressed helmet may not only entail a purchase at a good price, but also an investment in experience and knowledge.

All military headgear that has been around for more than 50 years, or which has actually seen battle service, deserves to be preserved irrespective of its appearance. We collectors have the duty to salvage and preserve all headgear, especially the most fragile examples. The author is one of the very few collectors who prefer to own worn headgear that shows heavy signs of use to the mint, unissued pieces.

Dirty, uncleaned headgear offers an opportunity to collectors. A close inspection of untouched sleepers witness the distinctive signs of age: a characteristic blue-grey patina on armour; crazing shrinkage on beavor; congealing on Picklehaubes; mothing and rips on felt; and hair loss on furs and animal manes. Some of the close up shots in this book illustrate these phenomena more or less clearly. It is also very interesting to study the manner helmet plates and fittings 'settled' on these distressed pieces. Removing a helmet plate from an unrestored Shako will frequently reveal a stain in the shape of the plate on the Shako body. Similar signs of oxidisation are noticed wherever metal is placed in contact with textile. A careful inspection of liners and chinstraps is equally

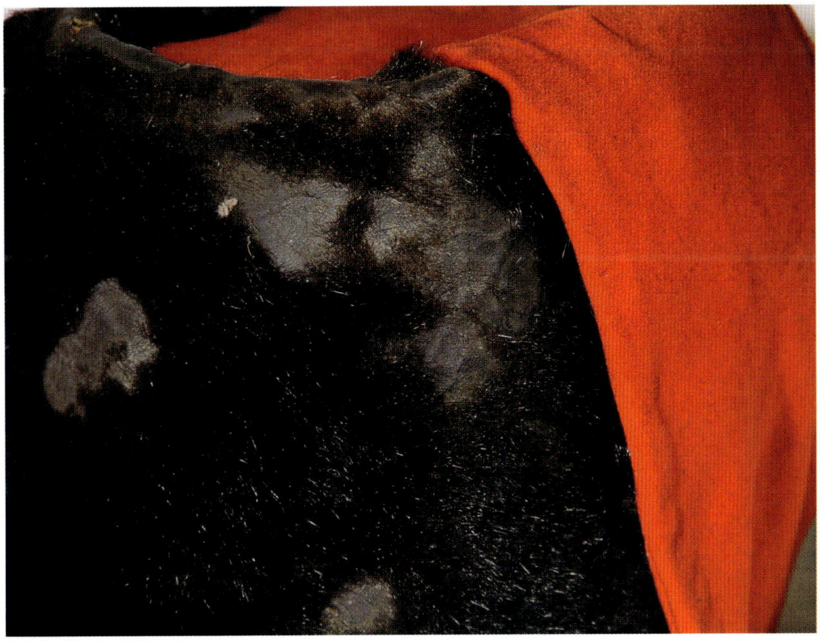

A British Busby c. 1880 showing characteristic signs of hair loss. (Private collection - Malta) Photo by M Bonnici

The top of a French WWI officer's Kepi. Note the high quality of stitching and signs of wear on the textile. (Private collection - Malta) Photo by M Bonnici

This rare original Crimean War era Russian Picklehaube appears to have been re-issued during its term of service.The front and back peaks appear to be old replacements. (Private collection - Malta) Photo by M Bonnici

intriguing. I noticed sweat stains and signs of wear on liners of helmets that are well over a hundred years old. Chinstraps often look dried and somewhat over stretched. Only collectors who have handled original items will recognise these signs of authenticity.

It is also important to keep in mind that some of the examples that are rated as being in a 'mint condition' or in a 'very good condition' are effectively original, but refurbished helmets. Leather is artificially reinvigorated with Vaseline and Petroleum Jelly, felts and beavers are dry-cleaned quasi-brand new. Picklehaubes are often expertly re-lacquered. Moth holes on Shakos and Kepis are closed with shavings from original textile. Blotches of hair loss on fur busbies and bearskins are sealed with glued patches of original cheap rabbit furs, and occasionally from expensive original bearskin recovered from British army surplus. Horse manes, woollen crest (Raups), and plumes are frequently replaced, because horsehair tends to fall with exposure to the elements. It is very important to know what you are buying, and I consider the acquisition of headdress in fair condition as a manner of avoiding occasional unpleasant surprises and buying at reasonable prices.

A Note on Spare Parts and Re-issues

Helmets have histories. Most of the headgear described in this book has been in circulation for over 75 years, which means that it has been subjected to a significant amount of age stress. Cheek plates from 16th century burgonets became loose or were lost, and were replaced way back in the 1850s. Visors and back peaks from 19th century Picklehaubes collapsed and were replaced with leather straps recovered from old leather marching boots. Burgonets were reconstructed from spare parts from armets and closed helmets. Some of our best headgear is not, strictly speaking, 100% original.

The world of antique military headgear is a thrifty one. Collectors of leather helmets from the first half of the 19th century notice arsenal plugs on their helmets, indicating that their helmet was re-issued at some point in time. Regimental helmet plates were substituted with re-organisations, and helmet bodies were recycled in view of shortages in supplies. Some French Shakos saw service under at least two regimes, bearing both a plate au coque and a plate a l'aigle during their term of service, while the Prussians cut down their Picklehaubes to ensure compliance with changing dress regulations. Some of our helmets have repairs that trace back to the headgear's period of active service, while others have more recent repairs.

Most advanced collectors and some dealers are well stocked with spare parts for antique military helmets; such as chin-scales, helmet plates, Shako pom poms, leather peaks, helmet plates, burgonet and Zischagge plates, spikes, and whatever is necessary to 'complete' a helmet. These original spare parts, acquired at very low cost, are being used to reconstruct helmets ex novo and to upgrade distressed helmets to a good condition. Honest sellers market these refurbished original helmets as composite helmets, but not all sellers are so honest. Composite helmets are identified with great difficulty, as it may be occasionally impossible to tell when a replacement or a repair was made. All the individual components of a helmet should be carefully examined before a purchase. Brims and peaks on Picklehaubes should be inspected to determine whether they fit well within the metal binding. Stitching should be examined under a magnifying glass, helmet plates should be dismounted, and extra holes (even if plugged) should be searched for. The collector should be aware that now his life has become more difficult than ever, and all his actions should be guided by sceptiscm and caution.

An exceptionally rare Model 1842 Prussian Reservists Private Purchase Picklehaube. (Private Collection Malta) Photo R. Attard

Rear view of the Model 1842 Picklehaube. The helmet is made of a light metal body covered with a a thin oilcloth cover simulating leather. (Private Collection Malta) Photo R. Attard

Bibliography

Modern Publications

Bull, Stephen: A Historical Guide to Arms … Armour (New York 1991)
Blaire Claude: European Armour (London 1958)
Boccia LG: Armi Bianche Italiani (Mila 1975)
Carman Michael: Head dresses of the British Army (Cavalry) (Surrey 1968)
Carman, W, Rowe: Head-dress of the British Lancers (1816-to the present) (Atglen 2002)
Cimarelli Aldo: Arms and Armour (London 1973)
Edge D, Paddock, M,: Arms and Armour of the Medeival Knight (London 1990)
Johansson Eric: Picklehauben (Missouri 1982)
Kube Jan: Study of German Uniforms and Headgear 1729-1918 (Atglen 1990)
Margerand Joseph: Les Coiffures de L' Armee' Francaise (Paris 1953)
Marzetti Paolo: Combat Helmets of the World (Parma 1996)
Muller Heinrich: Europaische Helme (East Germany-date of publication unknown)
Rowe David: Head-dress of the British Heavy Cavalry (Atglen 1999)
Steine, G, Moore I: The International Military Collectors Guide (London 1999)
Tavard, Christian: Casques et Coiffures Militaires Francais De L'Epoque Gauloise a la Second Guerre Mondiale (Paris 1981)
Rankin Robert: Military Headdress (London 1976)
Soares Branco Pedro: Coberturas Militares Portuguesas 1740-1918 (Lisbon-date of publication unknown)
Waverly Lewis: U.S. Military Headgear (1770-1880)
Wilkinson-Latham R.J: Collecting Militaria (New York 1975)

Some of the best publications on military helmets. Some have been long out of print. (Private collection - Malta) Photo by M Bonnici

Books (Antiquarian)

Bosio Giacomo: Della Sacra Religione…(First published in the 17th century)

Balbi Da Correggio: Diary of the Great Siege of Malta (First published in 1568)

Dal Pozzo Bartolomeo: Storia della…(First published in the eighteenth century)

Libr. Ms. 466 (Compiled in 1774)

Laking Guy Francis: A Record of European Arms and Armour (London 1922)

Newspapers

Gleason's Pictorial (Various Issues)

The Illustrated London News (Various Issues)

The Malta Observer (Various Issues)

The Graphic (Various Issues)

Catalogues/Price Guides etc

Alan Beadle Catalogues (Various)

Regimental's Catalogues (Various)

Sotheby's catalogues (Various issues)

Wallis and Wallis Catalogues (various)

Richard Austin: Military Collectibles

The Lyle Price Guide to Militaria Arms and Armour

Andrew Bottomley Catalogues (various)

The International Militaria Collectors Guide: Gary Sterne … Irene Moore

Spink Goiris-Voerhoef Collection of Militaria Catalogue November 1998

Kent Sales Catologues.

Other

The Bloomsbury Dictionary of Quotations

Notes

Disclaimer: